Poly Organo Phosphazene Nanomaterials for Therapy

by

S. Mehnath

Table of Contents

Chapter 1 – Introduction

1.1.   Introduction   1

1.2.   Poly[(organo)phosphazenes]   1
       1.2.1. Synthesis process   1
1.3.   Poly[bis(carboxyphenoxy)phosphazene]   2
       1.3.1. Functionalization of PCPP   3
       1.3.2. Drug loaded PCPP polymeric nanomaterial   4
       1.3.3. Suitability of PCPP   4
1.4.   Therapeutic drawbacks   5
       1.4.1. Cancer   5
       1.4.2. Tuberculosis   6
       1.4.3. Bone tissue engineering   7
1.5.   Scope of multifunctional polymeric nanomaterials   7
       1.5.1. Core-shell polymeric nanomaterial   8
       1.5.2. Hybrid Polymer   9
       1.5.3. Polymeric nanomaterial targeted drug delivery   9
       1.5.4. Nanofiber   10
       1.5.5. Alternative therapy   11
       1.5.6. Polymeric scaffold   12
       1.5.7. Polymer as coating material   13
1.6.   Objective of the thesis   14

Chapter 2 - Experimental Section

2.1.   Synthesis of poly[di(sodium carboxyphenoxy)phosphazene]   16
2.2.   Synthesis of Poly(bis (carboxyphenoxy) phosphazene)   17
2.3.   Preparation of drug-loaded polymeric nanomaterial   17
2.4.   Characterization techniques   17
       2.4.1. Nuclear Magnetic Resonance   17
       2.4.2. Fourier Transform-Infrared Spectroscopy   17
       2.4.3. X-ray diffraction   18
       2.4.4. Dynamic Light Scattering and Zeta potential measurement   18
       2.4.5. Thermal Gravimetric Analysis   18
       2.4.6. Scanning electron microscopy with energy-dispersive X-ray analysis   18
       2.4.7. Transmission Electron Microscope   19
       2.4.8. Atomic Force Microscope   19
       2.4.9. Compressive strength   19
2.5.   Physiochemical characterization   19
       2.5.1. Buffer Preparation   19
       2.5.2. Swelling studies   20
       2.5.3. Weight loss   20
       2.5.4. Biomineralization test   20
2.6.   Polymeric nanomaterials-drug interaction studies   20
       2.6.1. Evaluation of drug encapsulation efficiency   20

2.6.2. Drug release studies 21
2.7. *In vitro* anticancer studies 21
 2.7.1. Cell culture 21
 2.7.2. Cellular uptake 21
 2.7.3. Cytotoxicity assay of cancer cells 22
 2.7.4. AO/EB staining 22
 2.7.5. Hoechst Staining 22
 2.7.6. Determination of Reactive Oxygen Species 22
 2.7.7. Measurement of Mitochondrial Membrane Potential 23
 2.7.8. Western blotting 23
2.8. *In vitro* antibacterial studies 23
 2.8.1. Cell culture 23
 2.8.2. Bacterial culture 23
 2.8.3. Disc-diffusion studies 24
 2.8.4. Antimycobacterial effect of nanomaterials 24
2.9. *In vitro* bone regenerative studies 24
 2.9.1. Cell culture 24
 2.9.2. Cell proliferation 24
 2.9.3. Clonogenic assay 25
 2.9.4. Flow cytometry 25
 2.9.5. Osteoblast cells adhesion on TSS-HAP/PCPP3/PCL scaffold 25
 2.9.6. Statistical analysis 25

Chapter 3 - Co-encapsulation of dual drug loaded in MLNPs: Implication on sustained drug release and effectively inducing apoptosis in oral carcinoma cells

3.1. Introduction 26
3.2. Previous work done 26
3.3. Novelty 26
3.4. Objective 27
3.5. Experimental techniques 27
 3.5.1. Preparation of drug-loaded CCNPs coated by multilayer polymer 27
 3.5.2. FITC loaded MLNPs 28
 3.5.3. Animals 28
 3.5.4. *In vivo* antitumor effect of MLNPs 28
 3.5.5. Histopathological analysis 28
3.6. Results and discussion 29
 3.6.1. Characterization of PDCPP 29
 3.6.2. Synthesis and characterization of MLNPs 30
  (i) FTIR analysis 30
  (ii) DLS and Zeta potential 31
  (iii) SEM and TEM analysis 32
 3.6.3. Drug loading and encapsulation efficiency studies 34
 3.6.4. Drug release studies 35
 3.6.5. *In vitro* cancer studies 36
  (i) MLNPs cytotoxicity studies 36
  (ii) MLNPs cancer cell penetration studies 37
  (iii) AO/EB staining 38
  (iv) Hoechst staining 38
  (v) Determination of Reactive Oxygen Species 39
  (vi) Measurement of Mitochondrial Membrane Potential 40
 3.6.6. Western blotting 41
 3.6.7. *In vivo* antitumor effect of MLNPs 42
  (i) *In vivo* model body weight measurement 42

(ii) *In vivo* tumor regression studies   43
(iii) Biocompatibility of MLNPs using histopathology studies   45
3.7.   Conclusion   45

Chapter 4 - Thermoresponsive and pH triggered drug release of cholate functionalized poly(organophosphazene) polylactic acid copolymeric nanostructure

4.1.   Introduction   47
4.2.   Previous work done   47
4.3.   Novelty   47
4.4.   Objective   47
4.5.   Experimental techniques   48
   4.5.1. Preparation of PCPP-PLA-cholic acid   48
   4.5.2. Synthesis of Drug loaded hybrid polymeric nanomaterials   48
4.6.   Results and discussion   48
   4.6.1. NMR analysis   48
   4.6.2. FTIR analysis   50
   4.6.3. SEM analysis   51
   4.6.4. TEM analysis   52
   4.6.5. Thermoresponsive studies   53
   4.6.6. Swelling behavior   54
   4.6.7. *In vitro* drug loading and release   55
   4.6.8. Cytotoxicity assay   56
      (i) MTT assay   56
      (ii) Apoptosis assay   57
      (iii) Measurement of Mitochondrial Membrane Potential   58
   4.6.9. Western blotting   59
4.7.   Conclusions   60

Chapter 5 - Development of cholate conjugated hybrid polymeric nanomaterials for FXR receptor mediated effective site-specific delivery of paclitaxel

5.1.   Introduction   61
5.2.   Previous work done   61
5.3.   Novelty   61
5.4.   Objective   61
5.5.   Experimental techniques   62
   5.5.1. Synthesis of CA grafted PCPP conjugates   62
   5.5.2. Preparation of CA functionalized PDADMAC   62
   5.5.3. Preparation of CA-PCPP-PDADMAC-CA nanomaterials   62
   5.5.4. Comet Assay   63
   5.5.5. γH2AX Assay   63
5.6.   Results and discussion   63
   5.6.1. Preparation of CA-PCPP-PDADMAC-CA nanomaterials   63
   5.6.2. Characterization of CA-PCPP-PDADMAC-CA   64
      (i) FTIR analysis   64
      (ii)XRD pattern   65
   (iii) TGA studies   66
   (iv) DLS analysis   67
   (v) Zeta potential studies   68
   (vi) SEM analysis   69
   (vii) AFM analysis   70
   5.6.3. Swelling studies   71
   5.6.4. Encapsulation and *in vitro* drug release   72
   5.6.5. Cytotoxicity assay   74
   5.6.6. γH2AX assay   75

5.6.7. Comet assay   76

5.6.8. Western blotting   77

5.7.   Conclusion   78

Chapter 6 - Localized delivery of active targeting nanomaterials from nanofibers patch for effective breast cancer therapy

6.1.   Introduction   80

6.2.   Previous work   80

6.3.   Novelty   80

6.4.   Objective   81

6.5.   Experimental techniques   81

6.5.1. Preparation of PTX loaded PCPP-CA nanomaterial   81

6.5.2. PHM extraction and electrospinning solution preparation   81

6.5.3. Electrospinning of PTX@PCPP-CA in PHM nanofibers   82

6.5.4. *Ex vivo* skin permeation and drug release studies   82

6.5.5. Confocal microscopy analysis   82

6.5.6. Cytotoxicity assays   83

6.5.7. Cell morphology   83

6.6.   Result and discussion   83

6.6.1. Characterization of the PCPP-CA nanomaterial in nanofibers matrix   83

(i)   PTX@PCPP-CA nanomaterial SEM analysis   83

(ii)   PHM nanofibers SEM analysis   84

(iii)   PTX loaded PCPP-CA-PHM nanofibers SEM analysis   85

(iv)   AFM analysis   87

(v)   FTIR analysis   88

(vi)   XRD pattern   89

(vii) TGA studies   90

6.6.2. Swelling studies   91

6.6.3. Weight loss analysis   92

6.6.4. *In vitro* drug release   93

6.6.5. *Ex vivo* skin permeation study   94

6.6.6. Confocal microscopical analysis   95

6.6.7. TEM analysis   96

6.6.8. Cytotoxicity assays   97

6.6.9. Flow cytometry   98

6.6.10. SEM analysis   99

6.7.   Conclusion   100

Chapter 7 - Immunomodulating polyorganophosphazene-arginine layered liposome antibiotic delivery vehicle against pulmonary tuberculosis

7.1.   Introduction   101

7.2.   Previous Work   101

7.3.   Novelty   101

7.4.   Objective   102

7.5.   Experimental techniques   102

7.5.1. Synthesis of PCPP-*g*-arginine conjugates   102

7.5.2 Preparation of blank and dual drug-loaded liposome   102

7.5.3. Arginine-g-PCPP coated liposomes (PGL) preparation   103

7.6.   Results and discussion   103

7.6.1. Characterization of liposomes and PGL   104

(i) DLS and zeta potential of liposomes formulation   104

(ii) FTIR analysis ... 106
(iii) XRD pattern ... 107
(iv) TGA studies ... 108
(v) SEM analysis ... 109
(vi) TEM Analysis ... 112
7.6.2. Encapsulation and loading efficiency ... 113
7.6.3. *In vitro* drug release ... 114
7.6.4. Antimicrobial activity ... 115
7.6.5. Antimycobacterial activity ... 117
7.6.6. *In vitro* biocompatibility studies ... 118
7.6.7. Macrophage cell proliferation ... 119
7.7. Conclusion ... 120

Chapter 8 - Polyorganophosphazene combined rGO-MoS$_2$ nanomaterial with photo/chemotherapy for tuberculosis

8.1. Introduction ... 121
8.2. Previous work ... 121
8.3. Novelty ... 122
8.4. Objective ... 122
8.5. Experimental techniques ... 122
8.5.1. Synthesis of graphene oxide ... 122
8.5.2. Preparation of rGO-MoS$_2$ ... 122
8.5.3. Functionalization of PCPP with IZN ... 123
8.5.4. Fabrication of rGO-MoS$_2$-PCPP-IZN ... 123
8.5.5. ICG loading into rGO-MoS$_2$-PCPP-IZN ... 123
8.5.6. *In vitro* photothermal and photodynamic effect ... 123
8.5.7. Antibacterial effect of rGO-MoS$_2$-PCPP-IZN-ICG ... 124
8.5.8. Antimycobacterial effect of rGO-MoS$_2$-PCPP-IZN-ICG ... 124
8.5.9. Morphological characterization of *Mycobacterium* after irradiation ... 124
8.6. Results and discussion ... 124
8.6.1. Characterization of rGO-MoS$_2$-PCPP-IZN-ICG ... 125
(i) FTIR analysis ... 125
(ii) XRD pattern ... 125
(iii) TGA studies ... 126
(iv) SEM analysis ... 127
(v) TEM analysis ... 128
(vi) FE-SEM analysis ... 129
8.6.2. ICG loading ... 130
8.6.3. ICG stability ... 131
8.6.4. *In vitro* IZN and ICG release ... 132
8.6.5. *In vitro* photothermal effect ... 133
8.6.6. *In vitro* PDT studies ... 134
8.6.7. Antibacterial studies ... 135
8.6.8. Bacterial inhibition studies ... 136
8.6.9. Antimycobacterial activity ... 137
8.6.10. Flow cytometry ... 138
8.6.11. SEM analysis ... 139
8.6.12. Cytotoxicity assay ... 140
8.7. Conclusion ... 141

Chapter 9 - Bio-mediated hydroxyapatite coated 3D-PCPP polymer scaffolds for bone regeneration

9.1. Introduction ... 142

9.2.   Previous work   142
9.3.   Novelty   143
9.4.   Objective   143
9.5.   Experimental techniques   143
    9.5.1. Preparation of PCPP/PCL porous scaffold   143
    9.5.2. Synthesis of hydroxyapatite from shell   144
    9.5.3. Fabrication of HAP coated PCPP/PCL scaffold   144
    9.5.4. Scaffold porosity measurement   144
9.6.   Results and discussion   144
    9.6.1. Characterization of PCPP/PCL scaffold   145
        (i) FTIR and XRD analysis   145
        (ii) Porosity and mechanical strength of scaffold   145
        (iii) SEM analysis   146
    9.6.2. Characterization of natural shell HAP   147
        (i) FTIR analysis   147
        (ii) XRD pattern   147
        (iii) SEM and EDAX of TSS-HAP   148
    9.6.3. Characterization of HAP coated PCPP/PCL scaffold   149
        (i) FTIR analysis   149
        (ii) XRD pattern   149
        (iii) SEM and Elemental mapping   150
    9.6.4. Porosity, mechanical properties of scaffold   152
    9.6.5. TGA studies   153
    9.6.6. Weight loss of TSS-HAP-PCPP3/PCL scaffold   154
    9.6.7. Biomineralization test   155
    9.6.8. Cell proliferation and viability studies   156
    9.6.9. ALP activity of scaffold   157
    9.6.10. Clonogenic assay of scaffold   158
    9.6.11. MG-63 cell morophological studies   159
    9.6.12. Biocompatibility analysis of scaffold via flow cytometry   160
    9.6.13. SEM analysis   161
    9.6.14. Antibacterial activity of scaffold   162
9.7.   Conclusion   163

Chapter 10 - Coating of Ti-6Al-4V screw using mineralized marine sponge-PCPP polymer-HNT nanocomposite via electrophoretic deposition

10.1.   Introduction   164
10.2.   Previous work   164
10.3.   Novelty   165
10.4.   Objective   165
10.5.   Experimental techniques   165
    10.5.1. Preparation of Marine sponge loaded HNT   165
    10.5.2. Functionalization of MS loaded HNT with APTES   165
    10.5.3. Fabrication of PCPP grafted MS loaded HNT   166
    10.5.4. Titanium specimen surface pre-treatment   166
    10.5.5. Electrophoretic deposition   166
    10.5.6. Corrosion resistance behavior   166
    10.5.7. Bioactive agent release studies   166
10.6.   Results and discussion   167
    10.6.1.Charecterization of marine sponge-loaded HNT, polymer coated HNT   167
        (i) FTIR analysis   167

    (ii) XRD pattern 168
    (iii) TGA studies 170
    (iv) Zeta potential analysis 171
  10.6.2. MS loading of HNT 172
  10.6.3. MS release from HNT in different pH condition 172
  10.6.4. SEM analysis 173
  10.6.5. EPD and coating characterization of PCPP-HNT on Ti-6Al-4V screw 174
  10.6.6. Corrosion resistance studies 175
  10.6.7. SEM image of PCPP-HNT-MS coatings on Ti-6Al-4V screw 176
  10.6.8. SEM image of PCPP-HNT-MS apatite 177
  10.6.9. Cellular studies 180
    (i) Cell proliferation 180
    (ii) ALP activity 180
    (iii) Clonogenic assay 181
    (iv) MG-63 cells morophological studies 182
    (v) Biocompatibility effect of PCPP-HNT-MS via flow cytometry 183
  10.6.11. MG-63 cells adhesion on MS-PCPP-HNT 184
  10.6.12. Antibacterial activity 185
10.7. Conclusion 186
  Summary and Future scope of the work 187

Glossary of Abbreviations

PCPP	:	Poly[bis(carboxyphenoxy)phosphazene]
PDCPP	:	Poly[di(sodium carboxyphenoxy)phosphazene]
EDC	:	1-Ethyl-3-[3-dimethylami-nopropyl]-carbodiimide
NHS	:	N-hydroxysuccinimide
TB	:	Tuberculosis
WHO	:	World Health Organization
PTT	:	Photothermal therapy
PDT	:	Photodynamic therapy
ROS	:	Reactive oxygen species
ECM	:	Extracellular matrix
PTX	:	Paclitaxel
rGO	:	Reduced Graphene Oxide
MoS_2	:	Molybdenum disulphide
ICG	:	Indocyanine green
DCM	:	Dichloromethane
NMR	:	Nuclear magnetic resonance
FTIR	:	Fourier Transform-Infrared Spectroscopy
XRD	:	X-ray diffraction
JCPDS	:	Joint Committee on Powder Diffraction Standards
TGA	:	Thermal Gravimetric Analysis
DLS	:	Dynamic Light Scattering
SEM	:	Scanning Electron microscopy
FE-SEM	:	Field Emission- Scanning Electron microscopy
TEM	:	Transmission Electron Microscope
EDAX	:	Energy Dispersive X-Ray Analysis
AFM	:	Atomic Force Microscope
MES	:	2-ethanesulfonic acid
PBS	:	Phosphate buffer saline
EE	:	Encapsulation efficiency
MTT	:	(3-(4,5-Dimethylthiazol-2-yl)-2,5-Diphenyltetrazolium Bromide)
AO/EB	:	Acridine orange/Ethidium bromide
DCFH-DA	:	Dichloro-dihydro-fluorescein diacetate
TBST	:	Tris-buffered saline
EPR	:	Enhanced the permeability and retention
PELs	:	Polyelectrolytes
CCNPs	:	$CaCO_3$ nanoparticles
MLNPs	:	Multilayer Nanoparticles
PDADMAC	:	Poly (diallyl dimethyl ammonium chloride)
DMBA	:	7,12-Dimethylbenz[a] anthracene
Rh-123	:	Rhodamine-123
PLA	:	Polylactic acid
CA	:	Cholic acid
DMSO	:	Dimethyl sulfoxide

MMP	:	Mitochondrial Membrane Potential
FXR	:	Farnesoid X receptor
PHM	:	Psyllium Husk Mucilage
PI	:	Propidium Iodine
FDA	:	Food and Drug Administration
UV-Vis	:	Ultraviolet-Visible Spectrophotometer
RIF	:	Rifampicin
IZN	:	isoniazid
PGL	:	Arginine-g-PCPP polymer grafted liposomes
NIR	:	Near-Infrared
RNO	:	N,N-dimethyl-4- nitrosoaniline
BTE	:	Bone tissue engineering
PCL	:	Poly(ε-caprolactone)
HAP	:	Hydroxyapatite
TSS	:	Tower snail shell
BFS	:	Bicoloured Fan shell
CS	:	Clamshell
ALP	:	Alkaline phosphatase
Ti	:	Titanium
EPD	:	Electrophoretic deposition
HNT	:	Halloysite nanotubes
MS	:	Marine sponge
CS	:	*Callyspongia diffusa*
APTES	:	(3-Aminopropyl)triethoxysilane

List of Figures

Figure. 1.1.   Structures of poly(dichlorophosphazene) and R, R'location for the addition of different substituents which provides the functional properties.

Figure. 1.2.   Synthesis of Poly[bis(carboxyphenoxy)phosphazene].

Figure. 1.3.   PCPP polymeric nanomaterial with pH responsive and thermoresponsive nature.

Figure. 1.4.   Multifunctional polymeric nanomaterials for therapeutical applications.

Figure. 1.5.   Preparation of multilayer nanoparticles with a high amount of drug and controlled drug release.

Figure. 1.6.   Targeted drug delivery by polymeric nanomaterials.

Figure. 1.7.   Transdermal drug delivery from polymeric nanofiber for breast cancer.

Figure. 1.8.   Alternative/combinational therapeutic strategy.

Figure. 1.9.   Effective application of polymer coating of Titanium substrate via EPD.

Figure. 3.1. ^{1}H NMR spectrum of Poly (di (sodium carboxyphenoxy) phosphazene) in DMSO with TMS as an internal standard at 300 MHz.

Figure. 3.2. ^{13}C NMR spectrum of Poly (di (sodium carboxyphenoxy) phosphazene) in DMSO at 300 MHz.

Figure. 3.3.   FTIR spectra of (A) CCNPs (B) Cisplatin loaded CCNPs (C) Chrysin loaded CCNPs (D) CCNPs coated with PDCPP layer (E) CCNPs coated with PDADMAC layer (F) Both drugs loaded MLNPs.

Figure. 3.4.   (A) Average hydrodynamic diameters of naked CCNPs as a function of adsorption of polyelectrolyte multilayer. The odd numbers represent - PDADMAC layer and even numbers represent PDCPP layer, respectively (number 0, which represents the naked CCNPs) (B) Zeta potential measurements of drug-loaded CCNPs and sequentially layered PELs CCNPs. The odd numbers represent - PDADMAC layer and even numbers represent PDCPP layer, respectively (number 0, which represents the drug-loaded CCNPs zeta values).

Figure. 3.5.   SEM images represent the surface morphology of cisplatin and chrysin-loaded CCNPs (A) CCNPs synthesis by 800 rpm mixing (B) CCNPs synthesis by 1400 rpm mixing (C) CCNPs synthesis by 2000 rpm mixing (D) Demonstrates the high magnification of CCNPs coated with 5 layers of PDADMAC/PDCPP (E) Demonstrates the high magnification of $CaCO_3$ NPs coated with 10 layers of PDADMAC/PDCPP.

Figure. 3.6.   HR-TEM images of $CaCO_3$ NPs coated with 5 layers of PDADMAC/PDCPP (A) 1 µm, (B) 500 nm (C) 100 nm scale bar.

Figure. 3.7.   Drug loading and encapsulation efficiency of MLNPs from the encapsulation process and the co-precipitation process, with mixing speed and reaction time (A) Cisplatin (B) Chrysin.

Figure. 3.8.   The cumulative release profile of (A) Cisplatin (B) Chrysin from the $CaCO_3$ coated with PDADMAC-PDCPP MLNPs in pH 4.0, 7.4.

Figure. 3.9.   Cytotoxicity analysis after exposure of cisplatin and chrysin-loaded MLNPs to KB cells for (A) 12 h (B) 24 h. All the data are expressed as the mean $\pm$SD of the three experiments with duplicate wells. $p<0.05$ compared with the control group.

Figure. 3.10.   Cellular uptake of FITC loaded MLNPs using KB cells (A) Control cells under Phase contrast microscope (B) After 4 h treatment FITC loaded MLNPs (C) After 8 h treatment FITC loaded MLNPs.

Figure. 3.11.   Apoptotic effect of cisplatin and chrysin-loaded MLNPs into KB cells was stained by AO/EB and observed under a fluorescence microscope (A) Control (B) Chrysin MLNPs (C) Cisplatin MLNPs (D) Dual drug loaded MLNPs for 24 h.

Figure. 3.12.   MLNPs effect on KB cells nuclei damage was observed by Fluorescence microscope using Hoechst staining (A) Contol (B) Chrysin MLNPs (C) Cisplatin MLNPs (D) Dual drug loaded MLNPs for 24 h.

Figure. 3.13.   Effect of ROS generation by cisplatin and chrysin-loaded MLNPs into KB cells observed under a fluorescence microscope (A) Control (B) Chrysin MLNPs (C) Cisplatin MLNPs (D) Dual drug loaded MLNPs for 24 h (E) Bar diagaram representation of DCF fluorescence intensity, which is proportional to the amount of ROS produced by the KB cells. The Data represent mean $\pm$ SD.

Figure. 3.14.   Effect of MMP by cisplatin and chrysin-loaded MLNPs into KB cells observed under the fluorescence microscope (A) control (B) Chrysin MLNPs (C) Cisplatin MLNPs (D) Dual drug loaded MLNPs for 24 h (E)

Bar diagaram representation of Rh-123 fluorescence intensity in decreasing manner, which resembles the mitochondrial membrane depolarization on KB cells. The Data represent mean ± SD.

Figure. 3.15.   The expression of Bcl-2, Bax, caspase-3 in KB cells after treatment with chrysin MLNPs (A) control (B) Chrysin MLNPs (C) Cisplatin MLNPs (D) Dual drug loaded MLNPs for 24 h was analyzed western blot, respectively. The whole KB cell extracts were prepared and analyzed by Western blot analysis using antibodies against BCL-2, Bax, and caspase-3. The same blots were stripped and re-probed with a β -actin antibody to show equal protein loading.

Figure. 3.16.   Haematoxylin and Eosin stained tissue sections (A) Control hamsters showing normal epithelium of buccal tissue (B) DMBA alone treated hamsters (C) MLNPs alone + DMBA treated animal tissue (D) Chrysin loaded MLNPs +DMBA treated hamster (E) Cisplatin loaded MLNPs+DMBA (F) Chrysin and cisplatin loaded MLNPs +DMBA treated hamster.

Figure. 4.1. ^{1}H NMR spectrum of poly [bis (aryloxy) phosphazene] and Poly (bis (carboxyphenoxy) phosphazene) in DMSO with TMS an internal standard at 300 MHz.

Figure. 4.2. ^{13}C NMR spectrum of poly [bis (aryloxy) phosphazene] and Poly (bis (carboxy phenoxy) phosphazene) in DMSO at 300 MHz.

Figure. 4.3   FTIR spectra of (A) PCPP (B) PLA (C) CA (D) PCPP-PLA-CA hybrid polymer.

Figure. 4.4.   SEM image (A, B, C) Unprocessed hybrid polymer (D, E, F) Hybrid polymeric nanomaterials prepared using oil in water emulsion method (G, H, I) Drug loaded hybrid polymeric nanomaterials.

Figure. 4.5.   HR- TEM image of PTX loaded hybrid polymeric nanomaterial (A) 2 μm, (B) 1 μm, (C) 200 nm.

Figure. 4.6.   The hybrid polymer solution showed reversible gelation behavior at a temperature between (A) 37 °C (B) 20 °C.

Figure. 4.7.   Swelling behavior CA-functionalized PCPP-PLA hybrid polymer in pH 4.0, 7.4, 9.0 at 37 °C.

Figure. 4.8. PTX loading and encapsulation efficiency of PCPP-PLA-CA.

Figure. 4.9. The cumulative release profile of PTX from the of CA-functionalized PCPP-PLA hybrid polymeric nanoparticles in pH 4.0, 7.4, 9.0 at 37 °C.

Figure. 4.10. Cytotoxicity analysis after exposure of PTX loaded PCPP-PLA-CA to MCF-7 cells for 24 h. All the data are expressed as the mean ± SD of the three experiments with duplicate wells.

Figure. 4.11. Apoptotic effect of PCPP-PLA-CA into MCF-7 cells observed under a fluorescence microscope (A) Control (B) PCPP-PLA-CA (shows early apoptotic stage) (C) PTX (few cells at the late apoptotic stage) (D) PTX loaded PCPP-PLA-CA treated cells and stained with AO/EB staining.

Figure. 4.12. Effect of mitochondrial membrane potential (A) Control (B) PCPP-PLA-CA (C) PTX (D) PTX loaded PCPP-PLA-CA treated cells and stained with Rh-123.

Figure. 4.13. The expression of Bcl-2, Bax, caspase-3 in MCF-7 cells (A) control (B) PCPP-PLA-CA (C) PTX (D) PTX loaded PCPP-PLA-CA nanomaterial for 24 h was analyzed western blot, respectively. The same blots were stripped and re-probed with a β -actin antibody to show equal protein loading.

Figure. 5.1. ^{1}H NMR spectrum of CA-PCPP in DMSO at 300 MHz.

Figure. 5.2. ^{13}C NMR spectrum of CA-PCPP in DMSO at 300 MHz.

Figure. 5.3. FTIR spectroscopy of (A) CA conjugated PCPP (B) Sodium cholate conjugated PDADMAC (C) Polymeric nanomaterials (D) PTX loaded polymeric nanomaterials.

Figure. 5.4. XRD pattern of CA conjugated PCPP, sodium cholate conjugated PDADMAC, polymeric nanomaterials.

Figure. 5.5. Thermogravimetric analysis curve of CA conjugated PCPP, sodium cholate conjugated PDADMAC, polymeric nanomaterials.

Figure. 5.6. (A) DLS histogram of CA-PCPP-PDADMAC-CA (B) PTX loaded CA-PCPP-PDADMAC.

Figure. 5.7.   Zeta potential of CA-PCPP-PDADMAC and PTX loaded CA-PCPP-PDADMAC under different pH.

Figure. 5.8.   SEM image of (A) CA-PCPP-PDADMAC-CA polymer (2 µm) (B) 1 µm (C) PTX loaded CA-PCPP-PDADMAC-CA (500 nm) (D) 200 nm.

Figure. 5.9.   AFM 3D images of (A) CA-PCPP-PDADMAC-CA polymeric nanomaterials (B) PTX drug loaded CA-PCPP-PDADMAC-CA polymeric nanomaterials.

Figure. 5.10.   Swelling behavior CA-PCPP-PDADMAC-CA polymer in pH 5.0 and 7.0 at 25 and 37 °C.

Figure. 5.11.   Different concentrations of PTX encapsulation in CA-PCPP-PDADMAC-CA.

Figure. 5.12.   The cumulative release profile of PTX loaded CA-PCPP-PDADMAC-CA polymeric nanomaterial in pH 3.0, 5.0, and 7.4. at 37 °C.

Figure. 5.13.   (A) Cytotoxicity of CA-PCPP-PDADMAC-CA with different concentrations on MCF-7 cells after 24 h incubation (B) MCF-7 cell viability after 24 h incubation of PTX and PTX loaded CA-PCPP-PDADMAC-CA with different dosage at pH 7.0, 5.0.

Figure. 5.14.   γH2AX assay of cells treated with CA-PCPP-PDADMAC-CA for 12 h and 24 h.

Figure. 5.15.   Comet assay images of cells treated by CA-PCPP-PDADMAC-CA (A) Control (B) 12 h (C) 24 h.

Figure. 5.16.   Expression of Bax, Bcl-2, caspase-3, and β-actin in MCF-7 cells after treatment with IC 50 concentration of PTX loaded CA-PCPP-PDADMAC-CA (A) Control (B) 12 h (C) 24 h was analyzed western blot, respectively. The same blots were stripped and re-probed with a β-actin antibody to show equal protein loading.

Figure. 6.1.   (A) SEM image representing morphology of PTX loaded PCPP-CA (B) DLS curve of PTX loaded PCPP-CA.

Figure. 6.2.   (A) SEM image of 20 % of PHM nanofibers (B) 25 % of PHM nanofibers morphology represents in SEM. Size range of (C) 20 % of PHM nanofibers (D) 25 % PHM nanofibers.

Figure. 6.3.   SEM image representing size and shape of PTX loaded PCPP-CA-PHM nanofibers with component ratio of nanomaterial and PHM (A) 2:3 (B) 3:3 (w/w) ratio.

Figure. 6.4.   SEM image representing the shape of 1:3 (w/w) ratio PTX-PCPP-CA-PHM nanofibers on the different region (A) 10 μm (B) 10 μm (C) 10 μm (D) 5 μm.

Figure. 6.5.   Diameter of PTX-PCPP-CA-PHM nanofibers by DLS curve.

Figure. 6.6.   AFM image representing surface morphology of PTX-PCPP-CA-PHM nanofibers (A, C) 2D image (B, D) 3D image.

Figure. 6.7.   FTIR spectroscopy of (A) PCPP (B) CA (C) PCPP-CA (D) PHM (E) PCPP-CA-PHM nanofibers.

Figure. 6.8.   XRD pattern of (A) PCPP (B) CA (C) PCPP-CA (D) PHM (E) PCPP-CA-PHM nanofibers.

Figure. 6.9.   TGA curve analysis of PCPP, PCPP-CA, PCPP-CA-PHM nanofibers.

Figure. 6.10.   Swelling behavior of PTX-PCPP-CA-PHM nanofibers at pH 5.0, 6.0, 7.4.

Figure. 6.11.   Weight loss ability of PTX-PCPP-CA-PHM nanofibers at pH 5.0, 6.0, 7.4.

Figure. 6.12.   (A) Cumulative PTX release profile of PTX-PCPP-CA-PHM nanofibers from dialysis bag pH 5.0, 6.0, 7.4 (B) Nanofibers packed in a dialysis bag and immersed in buffer solution.

Figure. 6.13.   (A) *Ex vivo* skin permeation of PTX from nanofibers mat through Franz type diffusion cells pH 5.0, 6.0, 7.4 (B) Franz type diffusion cells apparatus based drug release through the skin membrane.

Figure. 6.14.   Internalization of FITC from nanofibers mat in skin section by Franz type diffusion cells apparatus at different pH (A, D) pH 5.0 (B, E) pH 6.0 (C, F) pH 7.0. Maximum green fluorescence permeation in pH 5.0 denotes the pH responsiveness of nanofibers.

Figure. 6.15.   PTX-PCPP-CA-PHM nanofibers degree of swelling were visualized by TEM image (A) Enlargement of nanofibers diameter, increases in void space between nanofibers (B) Images of core-shell nanofibers with nanomaterial (C) Presence of PTX-PCPP-CA nanomaterial inside core-shell nanofibers.

Figure. 6.16. (A) MCF-7 cell culture directly to PCPP-CA-PHM, PTX, PTX-PCPP-CA-PHM nanofibers viability was evaluated for days 1, 4, 8 (B) Nanofibers meshes were added, and fixed between two glass rings and viability was evaluated for days 1, 4, 8.

Figure. 6.17. Flow cytometry for identification of viable/dead (A) Quantitative analysis of cells without treatment and its histograms peak shift (B) Quantitative analysis of dead cells after PTX treatment and histograms peak shift (C) Quantitative analysis of dead cells after PTX-PCPP-CA-PHM nanofibers treatment and histograms peak shift.

Figure. 6.18. SEM images of MCF-7 cancer cells seeded over PTX-PCPP-CA-PHM nanofibers for (A, B) 12 h (C, D) 24 h maximum treatment time clearly showing the apoptotic body formation and membrane blebbing in treated cells.

Figure. 7.1. ^{1}H NMR Spectra of arginine-g-PCPP in DMSO at 300 MHz.

Figure. 7.2. DLS curve of (A) Liposomes (7:4) (B) Liposomes (7:6) (C) Liposomes (7:8) (D) Polymer grafted liposomes.

Figure. 7.3. FTIR spectroscopy of (A) Arginin-g-PCPP (B) Liposomes (C) Polymer grafted liposomes.

Figure. 7.4. XRD pattern of (A) Arginin-g-PCPP (B) Liposomes (C) Polymer grafted liposomes.

Figure. 7.5. TGA curve analysis of Arginin-g-liposomes, Liposomes, Polymer grafted liposomes.

Figure. 7.6. SEM image representing the structure and morphology of liposomes (A) 5 µm (B) 2 µm (C) 2 µm (D) 2 µm.

Figure. 7.7. SEM image representing the structure and morphology of liposomes prepared by lecithin/cholesterol (A) Molar ratio (7:6) (B) Molar ratio (7:8).

Figure. 7.8. (A,B) SEM image representing the shape of polymer grafted liposomes.

Figure. 7.9. TEM image of polymer grafted liposomes in different magnification (A-D) 100 nm (E-F) 50 nm.

Figure. 7.10. (A) *In vitro* cumulative drug release of RIF from liposomes and polymer grafted liposomes (B) *In vitro* cumulative drug release of IZN from liposomes and polymer grafted liposomes under different pH.

Figure. 7.11.   Antimicrobial effect of different strains of gram-positive and gram-negative bacteria against (A) RIF (B) IZN (C) RIF loaded PGL (D) IZN loaded PGL (E) RIF-IZN loaded PGL.

Figure. 7.12.   *Mycobacterium* growth inhibitory effect of RIF&IZN loaded PGL.

Figure. 7.13.   Biocompatible effect of RIF and IZN loaded PGL.

Figure. 7.14.   U937 macrophage cells proliferation effect of liposomes and polymer grafted liposomes in different concentrations.

Figure. 8.1.   FTIR of (A) GO (B) rGO-MoS$_2$ (C) PCPP-IZN (D) rGO-MoS$_2$-PCPP-IZN-ICG.

Figure. 8.2.   XRD of (A) GO (B) rGO- MoS$_2$ (C) PCPP-IZN (D) rGO-MoS$_2$-PCPP-IZN-ICG.

Figure. 8.3.   TGA of GO, rGO- MoS$_2$, PCPP-IZN, rGO-MoS$_2$-PCPP-IZN-ICG.

Figure. 8.4.   SEM image of GO (A) 5 µm (B, C) 3 µm, rGO-MoS$_2$ (D, E) 5 µm (F) 3 µm.

Figure. 8.5.   TEM image of GO (A) 0.5 µm (B) 200 nm (C, D) 100 nm (E) 50 nm (F) 20 nm.

Figure. 8.6.   TEM image of rGO-MoS$_2$ (A) 500 nm (B-D) 200 nm (E) 5 nm (F) 2 nm.

Figure. 8.7.   FE-SEM image of rGO-MoS$_2$-PCPP-IZN-ICG (A) 3 µm (B) 2 µm (C, D) 1 µm.

Figure. 8.8.   The UV-vis spectra of GO, rGO-MoS$_2$, PCPP-IZN, and rGO-MoS$_2$-PCPP-IZN-ICG.

Figure. 8.9.   UV-vis spectra of (A) Free ICG (B) rGO-MoS$_2$-PCPP-IZN-ICG stability compared by measured after 10 days.

Figure. 8.10.   (A) Cumulative IZN release from rGO-MoS$_2$-PCPP-IZN in different pH (B) ICG release from rGO-MoS$_2$-PCPP-IZN-ICG in different pH was measured by UV-vis spectra.

Figure. 8.11.   (A) *In vitro* photothermal effect of PCPP-IZN, rGO, rGO-MoS$_2$, ICG, rGO-MoS$_2$-PCPP-IZN-ICG (B,C,D,E) SEM image of rGO-MoS$_2$-PCPP-IZN-ICG after NIR-irradiation for 5 min.

Figure. 8.12.   (A) Singlet generation of rGO-MoS$_2$-PCPP-IZN without and with laser for 5 min (B) Singlet generation of rGO-MoS$_2$-PCPP-IZN-ICG without and with laser for 5 min.

Figure. 8.13.   Effect of rGO-MoS$_2$-PCPP-IZN-ICG inhibition in NIR-irradiation on (A) *E.coli* (B) *B.subtilis*.

Figure. 8.14.   Effective photothermal, photodynamic, antibacterial effect of rGO-MoS$_2$-PCPP-IZN-ICG after NIR-irradiation for different time interval and visualized by phase contrast microscope.

Figure. 8.15.   *Mycobacterium* growth inhibitory effect of rGO-MoS$_2$-PCPP-IZN-ICG in different concentration.

Figure. 8.16.   Flow cytometry for identification of viable/dead cells along with histograms peak shift of dead cells (A) ICG (B) rGO-MoS$_2$-PCPP-IZN (C) rGO-MoS$_2$-PCPP-IZN-ICG on *M.tuberculosis*.

Figure. 8.17.   (A-C) SEM images of *M.tuberculosis* without nanomaterial and NIR-irradiation (D-F) SEM images of *M.tuberculosis* after nanomaterial treatment and 5 min NIR-irradiation. Blue arrows indicate the cell damage and red arrows show the complete damage of *Mycobacterium* cell.

Figure. 8.18.   (A) Cytotoxicity of rGO-MoS$_2$-PCPP-IZN-ICG in different concentration (B) Comparison of cytotoxicity ability of nanomaterials.

Figure. 9.1.   FTIR, XRD image of (A, D) PCPP1/PCL scaffold, (B, E) PCPP2/PCL scaffold, (C, F) PCPP3/PCL scaffold.

Figure. 9.2.   SEM image of PCPP3/PCL scaffold in various magnification (A) 30 μm (B, C) 20 μm (D, E, F) 10 μm.

Figure. 9.3.   FTIR spectra of (A) HAP from *Turritella duplicate* (TSS-HAP), (B) HAP from *Pinna bicolor* (BFS-HAP) (C) HAP from *Mercenaria mercenaria* (CS-HAP).

Figure. 9.4.   XRD spectra of (A) HAP from *Turritella duplicate* (TSS-HAP), (B) HAP from *Pinna bicolor* (BFS-HAP), (C) HAP from *Mercenaria mercenaria* (CS-HAP).

Figure. 9.5.   (A,B) SEM image of HAP from *Turritella duplicate* (TSS-HAP) (C) EDAX analysis of HAP from *Turritella duplicate* (TSS-HAP).

Figure. 9.6.   FTIR of (A) TSS-HAP1PCPP3/PCL (B) TSS-HAP2PCPP3/PCL (C) TSS-HAP3PCPP3/PCL (D) TSS-HAP4PCPP3/PCL scaffold.

Figure. 9.7.   XRD image of (A) TSS-HAP1PCPP3/PCL (B) TSS-HAP2PCPP3/PCL (C) TSS-HAP3PCPP3/PCL (D) TSS-HAP4PCPP3/PCL scaffold.

Figure. 9.8.   SEM image of (A-D) TSS-HAP1PCPP3/PCL scaffold in different magnification (E-H) TSS-HAP4PCPP3/PCL scaffold in different magnification.

Figure. 9.9.   Elemental mapping of TSS-HAP4/PCPP3/PCL scaffold.

Figure. 9.10.   TGA analysis of (A) TSS-HAP (B) PCPP3/PCL scaffold (C) TSS-HAP1PCPP3/PCL scaffold (D) TSS-HAP2PCPP3/PCL scaffold (E)TSS-HAP3PCPP3/PCL scaffold (F)TSS-HAP4 PCPP3/PCL scaffold.

Figure. 9.11.   Weight loss of TSS-HAP1, TSS-HAP2, TSS-HAP3, TSS-HAP4 coated PCPP3/PCL scaffold.

Figure. 9.12.   (A-D) SEM image of TSS-HAP4/PCPP3/PCL scaffold after treatment with SBF for 7 days (E-H) SEM image of TSS-HAP4/PCPP3/PCL scaffold after treatment with SBF for 21 days.

Figure. 9.13.   MG-63 cell proliferation rate after treatment with different concentration of (A) TSS-HAP (B) TSS-HAP1PCPP3/PCL (C) TSS-HAP2PCPP3/PCL (D) TSS-HAP3PCPP3/PCL (E) TSS-HAP4PCPP3/PCL scaffold.

Figure. 9.14.   MG-63 cell ALP activity after treatment with different concentration of (A) TSS-HAP (B) TSS-HAP1PCPP3/PCL (C) TSS-HAP2PCPP3/PCL (D) TSS-HAP3PCPP3/PCL (E) TSS-HAP4PCPP3/PCL scaffold.

Figure. 9.15.   Clonogeny study of MG-63 cells after treatment with TSS-HAP4 coated PCPP3/PCL scaffold for 7, 14, 21 days.

Figure. 9.16.   Visualization of MG-63 cells morphology after treatment with TSS-HAP, PCPP3/PCL scaffold, TSS-HAP1, TSS-HAP2, TSS-HAP3, TSS-HAP4 coated PCPP3/PCL scaffold for 0 h, 12 h, 24 h by phase contrast microscope.

Figure. 9.17.   (A,B) PCPP3/PCL scaffold biocompatible effect on MG-63 cells (C,D)TSS-HAP4 coated PCPP3/PCL scaffold biocompatible effect on MG-63 cells (E,F) Positive control cells biocompatible effect measured by flow cytometry.

Figure. 9.18.   SEM image represents the MG-63 cells growth, adhesion over the TSS-HAP4 coated PCPP3/PCL scaffold.

Figure. 9.19.   Antiabcterial effect of PCPP3/PCL, TSS-HAP4, TSS-HAP4-PCPP3/PCL (A) *Staphylococcus aureus* (B) *Escherichia coli.*

Figure. 10.1.   FTIR micrographs of (A) 100 mg marine sponge loaded halloysite (B) 200 mg marine sponge loaded halloysite (C) 300 mg marine sponge loaded halloysite.

Figure. 10.2.   FTIR analysis of (A) HNT (B) AEAPS functionalized nanoclay (C) Polymer coated halloysite.

Figure. 10.3.   XRD pattern of (A) 100 mg marine sponge loaded halloysite (B) 200 mg marine sponge loaded halloysite (C) 300 mg marine sponge loaded halloysite.

Figure. 10.4.   XRD pattern of (A) HNT (B)AEAPS functionalized nanoclay (C) Polymer coated halloysite.

Figure. 10.5.   TGA analysis of nanoclay and PCPP-HNT.

Figure. 10.6.   MS release from PCPP functionalized HNT under different pH condition.

Figure. 10.7.   SEM image represents PCPP functionalized MS-HNT in different magnification

Figure. 10.8.   Potentiodynamic polarization curves of (A) Ti-6Al-4V screw PCPP-HNT-MS/voltage varied samples (B) 0.3/10 (C) 0.3/20 (D) 0.3/30 (E) 0.6/10 (F) 0.6/20 (G) 0.6/30 coated Ti-6Al-4V screw.

Figure. 10.9.   SEM image of PCPP functionalized MS-HNT coated titanium screw surface via EPD at different magnification.

Figure. 10.10.   SEM image of (A) PCPP functionalized HNT (B-F) Biomineralization PCPP functionalized HNT immersed in SBF for 21 days which visualized in SEM at different magnification.

Figure. 10.11.   EDS analysis of PCPP-HNT-MS before SBF treatment.

Figure. 10.12.   EDS analysis of PCPP-HNT-MS after SBF treatment.

Figure. 10.13.   Elemental mapping of PCPP functionalized HNT-MS after 21 days of SBF treatment.

Figure. 10.14. Cell proliferation ability of HNT, PCPP-HNT, MS-PCPP-HNT on MG-63 cells.

Figure. 10.15. ALP activity of HNT, PCPP-HNT, MS-PCPP-HNT on MG-63 cells.

Figure. 10.16. Clonogeny study of MG-63 cells after treatment with HNT, PCPP-HNT, MS-PCPP-HNT for 7, 14, 21 days.

Figure. 10.17. Visualization of MG-63 cells morphology after treatment with HNT, PCPP-HNT, MS-PCPP-HNT for day-0, 3, and 7 by phase contrast microscope.

Figure. 10.18. Flow cytometry for identification of viable/dead cells along with histograms peak shift of dead cells (A,B) HNT (C,D) PCPP-HNT (E,F) PCPP-HNT-MS on MG-63 cells.

Figure. 10.19. SEM image represents the MG-63 cells growth, adhesion over the MS-PCPP-HNT.

Figure. 10.20. Antiabcterial effect of HNT, PCPP-HNT, MS-PCPP-HNT (A) *Staphylococcus aureus* (B) *Escherichia coli.*

List of Table

Table. 2.1.   Specific λ-max value of different drug.

Table. 3.1.   Initial and final body weight changes of control and experimental hamsters in each group.

Table. 3.2.   Tumor incidence, tumor volume and tumor burden in the control and experimental animal in each groups (n=6).

Table. 7.1.   Composition of different liposomal formulations.

Table. 7.2.   Zeta potential value of different liposomal formulations.

Table. 7.3.   RIF and IZN drug loading and encapsulation efficiency of Polymer grafted liposomes.

Table. 7.4.   Antimicrobial effect of different strains of gram-positive and gram-negative bacteria against RIF, IZN, RIF loaded PGL, IZN loaded PGL, RIF-IZN loaded PGL, and its MIC, MBC and Zone of inhibition.

Table. 8.1.   ICG loading efficiency.

Table. 9.1.   Porosity and mechanical strength value of different concentration of PCPP/PCL scaffold.

Table. 9.2.   Porosity, mechanical properties TSS-HAP 1-4 coated PCPP3/PCL scaffold.

Table. 10.1.   Zeta potential value of HNT, MS-HNT, PCPP-HNT.

Table. 10.2.   MS loading, encapsulation in HNT and PCPP-HNT.

Table. 10.3.   Thickness and deposit weight of PCPP-HNT-MS on Ti-6Al-4V screw in different condition.

Abstract

A key purpose of the present work is to develop a different form Poly[(organo)phosphazenes] polymeric nanomaterials against variety of therapeutic applications like cancer, tuberculosis and bone-tissue engineering. To achieve the desired property, PCPP was combined with polymer, inorganic, organic, metal, liposomes, carbon-based materials and biomaterial. The addition of these components can enable specific functionalities like higher drug loading ability, stimuli-responsive drug delivery, site specific/*in-situ* delivery, tissue regeneration, implant coating, and alternative therapeutic modality. The enhanced superior properties of PCPP overcomes the limitation of conventional polymer drug delivery and resolve the many therapeutic defects.

Accordingly, this thesis comprises of ten chapters which are as follows:

Chapter 1   Introduction describes fundamental concepts of Poly[bis (carboxyphenoxy) phosphazene] polymeric nanocarrier, functionalization, fabrication into various forms like core-shell, hybrid polymer, liposomes, photoactive nanomaterials, implant, scaffold application on drug delivery. The mechanism for drug loading, release ability, interaction with other type of nanomaterials.

Chapter 2   This chapter deals briefly with the principles involved in various characterization techniques employed for identifying the properties of the fabricated coating materials and the respective forms of applications.

Chapter 3   This chapter deals with the effect of PDCPP as dual drug carrier with the help of another synthetic polymer PDADMAC via electrostatic interaction. MLNPs reports the physiochemical, *in vitro* and *in vivo* anticancer effect against oral cancer.

Chapter 4   This chapter gives a perception of hybrid polymer, PCPP-PLA-CA conjugates for stimuli responsive drug delivery against breast cancer. Hybrid polymer exhibits the temperature based modification of side chain and breakdown of ester linkage at acidic pH.

Chapter 5   The chapter gives an outlook of nanomaterial based targeted drug delivery of PTX against breast cancer. FXR targeting nanomaterial prepared using the PCPP and CA along with PDADMAC. It mainly focus on site-directed nuclear delivery of drug which implies towards successful cancer cell death.

Chapter 6   This chapter details on the development of electrospinning technique to fabricate PCPP nanomaterial loaded core-shell implantable nanofibers. The structural morphology, sustained drug release, *in-situ* delivery, skin permeation ability enhances the anticancer drug delivery effect.

Chapter 7   The chapter specifies the importance of targeted drug delivery and immunomodulation effect against tuberculosis. PCPP functionalized arginine was coated with liposomes for effective delivery of hydrophobic and hydrophilic drug to TB-infected alveolar macrophage.

Chapter 8   The chapter gives detailed role of PCPP polymer combination with photoactive nanoagent like rGO-MoS$_2$ and ICG. The nanomaterial causes the photothermal, photodynamic effect, along with delivery of antimycobacterial drug against *M.tuberculosis*.

Chapter 9   The chapter describes the effect of PCPP in scaffold preparation and evaluating the interaction of hydroxyapatite from different natural shell, towards orthopedic application.

Chapter 10   The chapter brief the PCPP interaction with nanoclay ceramic loaded with marine sponge minerals, and the nanocomposites was coated in titanium screw via electrophoretic deposition. The study confirms the application polymeric biomaterial ceramic.

Summary and Conclusions highlighting the results and major conclusions drawn during the present investigation.

1.1. Introduction

Previously, polymeric nanomaterial was used to protect the drugs and increase the pharmacokinetics of drugs during the delivery process. Later it gained a huge response from researchers and has become the key factor for multifunctional nanomedicine [1]. The main concept of multifunctional nanomaterial is referring to loading/delivery of more than one drug, tunable properties, adjusting to the environment, diagnosis, targeted, and controlled delivery of drugs. It is also applicable to the use of different delivery strategies and various therapeutic methods like chemotherapy, immunotherapy, and photothermal (PTT)/photodynamic therapy (PDT) [2-4]. The multifunctional nanomaterial should be biocompatible, biodegradable, highly stable, improves the hydrophilicity, bioavailability of the drug, control the drug interaction of cellular plasma/proteins, enhance the efficacy of therapeutic agent, and better accumulation of the drug in target site [5,6]. Additionally, it can conjugate with chemically diverse components like hydrophobic/hydrophilic drugs, natural/synthetic polymers, amino acids, targeting ligands, proteins, and antibodies, etc. Such functionalization makes the multifunctional nanomaterial with great potential to treat diverse diseases [7,8]. Developing a polymer with multifunctional properties requires the careful selection of polymer and needs to satisfy several demands before considered for use in diverse therapeutic applications.

1.2. Poly(organo)phosphazenes

Poly(organo)phosphazenes is an alternating phosphorus, nitrogen, and organic substituent-based hybrid polymer. The polydichlorophosphazene is a precursor, which is extremely unstable but can be easily substituted with nucleophilic substituents to get a various range of stable poly[(organo)phosphazenes]. In Figure.1.1, the -R groups offer the location for the addition of different substituents which provides the functional properties of the polymer. This flexibility resulted in the development of a variety of versatile polymers with a wide range of applications [9,10].

1.2.1. Synthesis process

The synthesis of poly(organo)phosphazenes was performed in a two-step process starting with the thermal ring-opening polymerization of hexachlorocyclotriphosphazene. The conversion of cyclic trimer to linear poly(dichlorophosphazene). Further, organic substituent was added to the polymer through nucleophilic macromolecular substitution for the phosphorus-bound chlorine atoms [11].

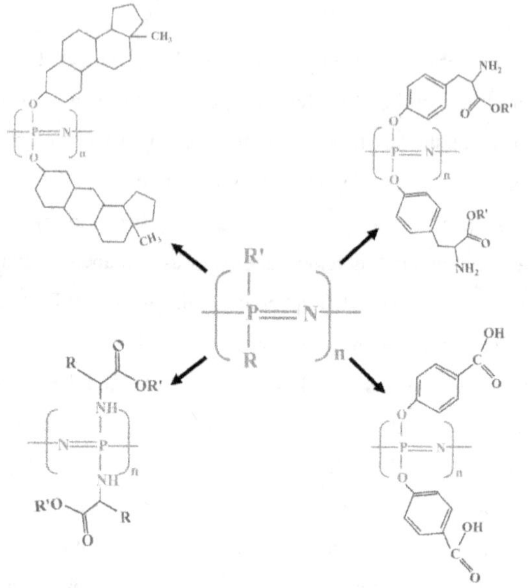

Figure.1.1. Structures of poly(dichlorophosphazene) and R, R'location for the addition of different substituents which provides the functional properties.

The thermal ring-opening polymerization process was attempted by H. N. Stokes during the late 1800s and his product was insoluble due to crosslinking. After that, Allcock and Kugel (1965) synthesized the linear poly(dichlorophosphazene) under well-controlled thermal ring-opening polymerization. The product was easily soluble and allowed further modification P–Cl bonds with organic nucleophiles. Each polymer has one type of side-chain substituent throughout the polymeric chain. The well-defined ratio of side-chain was added to afford the polymer with desired properties like stability, controlled degradation, and functionalization [12,13].

1.3. Poly[bis(carboxyphenoxy)phosphazene]

The Poly[bis(carboxyphenoxy)phosphazene] (PCPP) is a carboxylate, the important group of poly(organo)phosphazenes polymer. These phenoxy carboxylate polymers were in two forms free acid and sodium salt. PCPP was also prepared by synthesizing the poly(dichlorophosphazene) precursor by thermal bulk polymerization of hexachlorocyclotriphosphazene (Figure.1.2). Then, the precursor was reacted with propyl p-hydroxybenzoate to form poly[bis(aryloxy)phosphazene] ester. Here the chlorine atom was replaced by carboxylate ester-containing side groups. Finally, hydrolysis of the ester group to form a carboxylic acid group-containing PCPP [14,15]. This PCPP polymer was widely used for encapsulation of drugs/antibiotics, proteins, vaccines, and cells [16,17]. As with all poly[(organo)phosphazenes], PCPP can also be tailored to achieve desired properties. It is also exhibited in different forms like microsphere and hydrogel by substitution of divalent calcium salt [18].

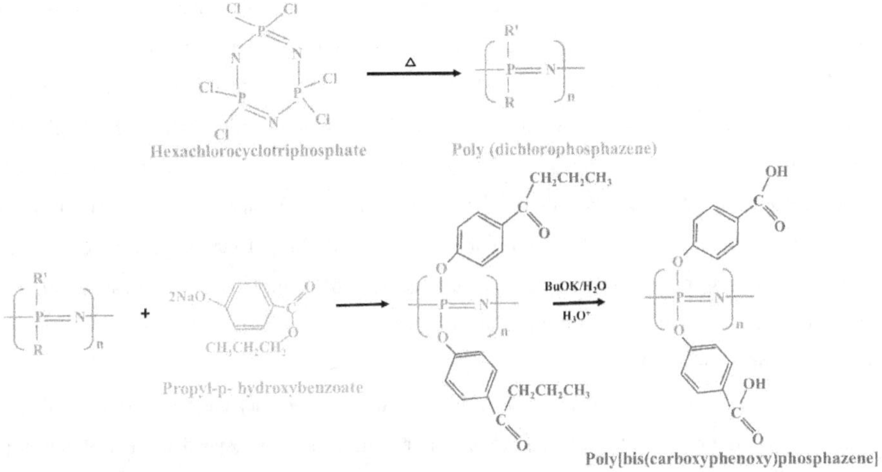

Figure.1.2. Synthesis of Poly[bis(carboxyphenoxy)phosphazene]

1.3.1. Functionalization of PCPP

The polymer functionalization with other polymers, ligand, amino acids, proteins, etc. was important to enhance the PCPP properties like targeted delivery, stimuli-responsive delivery. PCPP polymers consist of the carboxylic acid group and it easily functionalizes with other amino groups containing components via EDC/NHS reaction. PCPP functionalization was also performed with the help of an ethylene diamine crosslinking agent [19,20]. EDC-NHS reaction is well-known organic and bioconjugate coupling chemistry. In this, EDC (1-Ethyl-3-[3-dimethylami-nopropyl]-carbodiimide hydrochloride) is a cross-linking agent used to conjugate the carboxylic acid groups to the primary amines and it formed by amine-reactive NHS-esters. EDC activates the carboxylic groups and reacts to form an amine-reactive O-acyl isourea intermediate. This intermediate does not react with an amine group, it is unstable, hydrolyze to restore the carboxyl group. But, in the occurrence of N-hydroxysuccinimide (NHS), EDC converts the carboxylic group to amine-reactive NHS esters [21,22].

1.3.2. Drug loaded PCPP polymeric nanomaterial

There are several methods available for polymeric nanomaterial preparation like emulsion-solvent evaporation, nanoprecipitation, solvent diffusion, and phase inversion. For encapsulation of hydrophobic drugs/antibiotics, the most commonly used techniques are emulsion-solvent evaporation and nanoprecipitation. In emulsion-solvent evaporation method is mixing of non-water miscible solvent and an aqueous solution containing emulsifying agent with high shear force. The evaporation of volatile solvent forms polymeric nanomaterial. The advantage of this technique is the formation of smaller particles, non-toxic, and faster reaction. The drawbacks are standardization for every specific drug, higher energy affects the stability of drugs. The nanoprecipitation method is simple, requires lower energy, and good reproducible technique. It involves the utilization of miscible solvents and which are emulsified in an aqueous medium containing a surfactant. The polymeric nanomaterial formulation was purified from toxic impurities, organic solvent/surfactants, polymeric residues, and unreacted drugs were removed by a separate process. The majorly used purification process was dialysis, ultracentrifugation, gel-filtration, and evaporation under pressure [23,24].

1.3.3. Suitability of PCPP

PCPP polymer is compatible with biological environments and it acts as biodegradable material to form non-toxic degradation products like ammonia and phosphates. The different forms of drugs/antibiotics were shielded by the polymer's phosphorus-nitrogen backbone. The substituent of linkage in the PCPP polymer helps in tunable degradation rates and prevents uncontrollable leakage of drugs. PCPP alters its physical/chemical structure based on the stimuli like pH, temperature, reductive, oxidative, and enzyme reaction. Majorly, PCPP responds to multi-stimuli responsive like acidic pH and temperature (Figure.1.3). A widely explored ester linkage that responds to the acidic pH. The introduction of ester linkage into the PCPP polymer causes the breakdown and degradation of PCPP polymer in acidic pH. This causes a high release of drug from the PCPP and the diffusion rate was faster compared to the neutral/basic pH. The high drug retaining ability of PCPP in an acidic medium helps to release more drugs in the cell with an acidic environment [25,26].

Thermoresponsive properties of PCPP depend on the activation of hydrophilic and hydrophobic constituents based on the temperature. The sol-gel transformation of PCPP has a more advantage and it act as injectable for site-directed drug delivery. It forms an aqueous solution at 25 °C and spontaneously transforms into gel at 37 °C. Further it enhances the bioavailability of drugs at targeted sites by injecting them at the specific site. The formation of gel at body temperature (37 °C) leads to releases a drug in a sustained manner. The gel strength of PCPP was increased by modifying the chemical cross-linkage (eg. Thiol groups) [27].

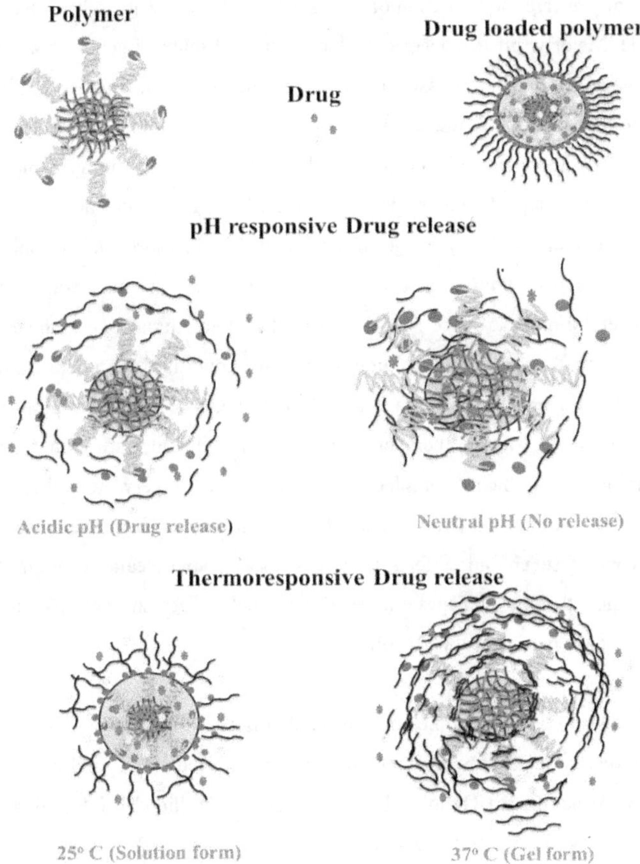

Figure.1.3. PCPP polymeric nanomaterial with pH responsive and thermoresponsive nature.

1.4. Therapeutic drawbacks

Recently, human health is susceptible to various diseases and new types of drugs must be discovered to resolve the critical issue. Every therapeutic challenge was varied based on the diseases.

1.4.1. Cancer

Currently, cancer is one of the leading causes of mortality worldwide. After China and the USA, India was the leading country for having new cancer cases every year (1 million/year) and half a million cancer deaths every year. In that 70 % of patients are women and breast cancer, cervical cancer, and oral cancer are most common in India. Breast cancer was majorly found among women of age 40. The symptoms are the formation of lumps, change in size/shape of the breast, blood discharge from the nipple. Majorly, breast cancer cells have inherited mutated genes breast cancer gene 1 and breast cancer gene 2 which was reason to cause the breast cancer. There was different staging for breast cancer formation, stage 0- ductal carcinoma development, stage 1- tumor grow upto 2 cm, stage 2- tumor grow across the 2 cm and spread to nearby nodes, stage 3- tumor grow upto 5 cm and spread to other lymph nodes, satge 4- tumor spread to bones, liver, brain, or lungs. The diagnosis tool is a mammogram, containing an X-ray image of the breast showing normal/abnormal tissues. Next, oral cancer is the third most common cancer in India, and tobacco, alcohol consumption is the major reason for cancer formation. Cancer developed in the mouth/throat tissue and the squamous cells present in the mouth, tongue, and lips region. In extreme condition, cancer spread through the lymph nodes of the neck region. The symptoms are chronic ulcers in the oral region, loss of teeth, and difficulty in swallowing. Mostly, therapeutic options for both cancers are chemotherapy (paclitaxel (PTX), doxorubicin, epirubicin, paraplatin, 5-fluorouracil), surgery, radiotherapy, and hormonal therapy are available to improve a cancer patient's life. The cancer chemotherapy antitumor agents are hydrophobic, causing severe side effects to normal tissue, rapid clearance during blood circulation. Similarly, radiation therapy damages the nearby normal tissue, chances of cancer relapsing, and hormone therapy have poor efficacy [28,29].

1.4.2. Tuberculosis

Tuberculosis (TB) is a chronic communicable disease caused by *Mycobacterium tuberculosis*. Approximately one-third of the world's population was infected by TB and caused over one million deaths annually. As per the WHO global TB report 2019, 1.7 million TB infected people died and 10.4 million new TB cases rise. Several effective antibiotics (rifampicin (RIF), isoniazid (IZN), ethambutol, and pyrazinamide) are available, yet TB persists as a major challenge owing to control and transmission [30]. The *Mycobacterium* localization inside the cells protects them from lysosomal attack, humoral, and cellular immune response, which confines the efficacy of antibiotic treatments. This leads to the utilization of multiple drugs for a long-time period to increase the availability of drugs [31]. But the contraindication of combinational drugs by serious side effects and intolerance leads to the rise of single or multidrug-resistant strains. WHO reported 2,50,000 multidrug-resistant-

TB deaths occurred in 2019. Six countries accounted for 60 % of the new cases where India leads the count followed by Indonesia, China, Nigeria, Pakistan, and South Africa [32,33].

1.4.3. Bone tissue engineering

Bone is a natural composite composed of collagen, carbonated apatite, and other non-collagenous proteins which helps in the stimulatory microenvironment for cellular functions. In general, bone can heal by itself with a multistage repair process. The osteoblasts (bone-forming cells) and osteoclasts (bone-resorbing cells) are important cells responsible for bone regeneration [34]. But it was difficult for self-healing of bone during the larger bone defects/fracture. In such cases, bone tissue engineering (BTE) plays a vital role in providing grafts, implants, or scaffolds for tissue regeneration [35]. The range of materials that are available to treat such problems includes autologous bone, allogeneic bone, and demineralized bone matrices. Transplantation of autogenous and allogeneic bone was a standard method for fixing bone defects but the application of these bone grafts is limited due to the inadequate supply, donor site injury, and potential infection [36]. So, artificial bone substitutes have great attention, and they are categorized as ceramics, calcium sulfate, bioglass, hydroxyapatite (HAP), and tricalcium phosphate. These substituents are lack in biocompatiblity, absent bioactive substances, minerals, proteins, etc. are the drawback [37].

1.5. Scope of multifunctional polymeric nanomaterials

To extend the therapeutic application, polymers can also be combined with metal, an inorganic, organic, carbon-based material, biomaterial, and ceramic, etc. Such combination can modify the properties of nanomaterial, pharmacokinetics of a drug, delivery strategy and allows the utilization of alternative therapeutic strategies like immunotherapy, PTT, and PDT. This leads to the development of a multifunctional system like core-shell polymer, hybrid polymers, polymeric micelles, nanomembrane, hydrogel, scaffold, polymeric implant, etc., such systems are often referred to as "polymer therapeutics" (Figure.1.4) [38,39].

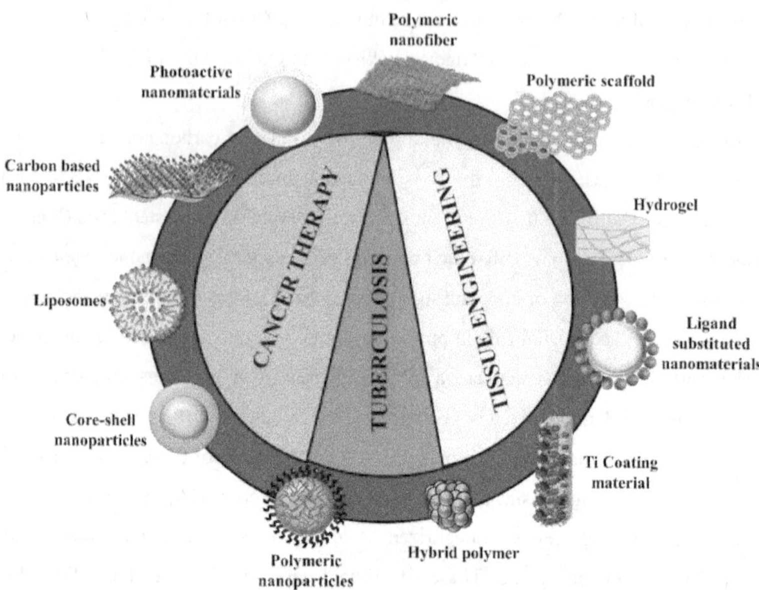

Figure.1.4. Multifunctional polymeric nanomaterials for therapeutical applications.

1.5.1. Core-shell polymeric nanomaterial

One of the most promising techniques for the fabrication of core-shell nanomaterial is the electrostatic Layer-by-Layer self-assembly technology. The simple method was discovered by Decher using a flat substrate and in the future, it extends for colloidal particles. Core-shell polymeric nanomaterial consists of two compartments, polyelectrolyte layer, and core wall. The polyelectrolyte layer was further added to form a multilayer via electrostatic interaction. It is formed by the consecutive deposition of oppositely charged polyelectrolytes (PELs) onto colloidal core templates. This high amount of drug can be loaded into the core and multi-layer helps in loading the maximum number of drugs (Figure.1.5). Core-shell can be tailored using conventional PELs for size, shape, and thickness. The higher surface interaction allows a controlled release of encapsulated drugs from the multi-layer nanosystem [40,41].

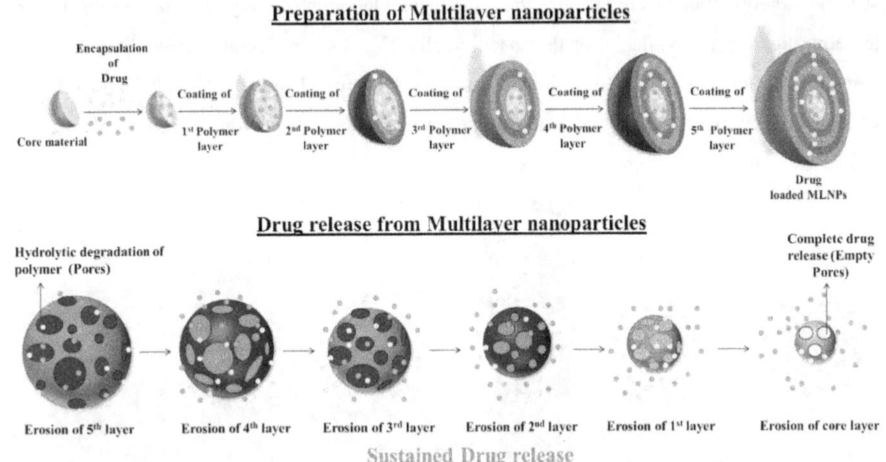

Figure. 1.5. Preparation of multilayer nanoparticles with a high amount of drug and controlled drug release.

1.5.2. Hybrid polymer

To achieve desirable degrading properties, drug release kinetics, and thermoresponsive properties from a single polymeric nanomaterial were achieved through a hybrid polymer. The combination of a hydrophobic and hydrophilic polymer with a specific molecular size can form a hybrid polymer. The inclusion of hydrophobic polymer into the hydrophilic PCPP polymer plays a vital role in pH-responsive degradation and effective gelation. Hydrophilic components help in water penetration and hydrophobic components would increase the porosity of the material. Hydration is caused by hydrophilic molecules which cause steady increases in water penetration. Simultaneously, hydrophobic polymer controls the higher swelling, burst drug release. The proper balance between those components makes desired polymeric nanomaterials [42,43].

1.5.3. Polymeric nanomaterial targeted drug delivery

In targeted drug delivery, drugs must reach their site of action and for that drug should have high permeability against different biological barriers. In several cases to achieve the high permeability, an excessive amount of drug was given to the patients and this leads to the extensive distribution of the drug in the body. This causes a severe side effect, accumulation in the other sites, and limits the pharmacological activity. In order to overcome this issue, cell-targeted drug delivery techniques were recommended. Here, the drug-carrying polymeric nanomaterial was decorated with specific targeting moieties to enhance the drug accumulation at the targeted site (Figure. 1.6). Several targeting ligands are available for the specific cells (Eg. small molecules, antibody, aptamers, folate, mannose, proteins, and antibody derivatives, etc.) [44-46]. Targeting ligand decorated polymeric nanomaterials delivers the drug at a specific site, prevents nonspecific delivery, and enhances the drug efficiency.

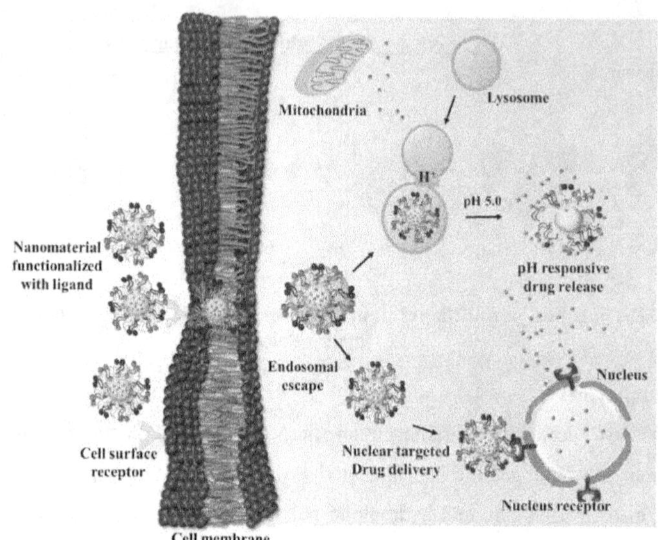

Figure. 1.6. Targeted drug delivery by polymeric nanomaterials.

1.5.4. Nanofiber

The delivery strategy is very important to improve drug efficiency, control side effects, reduce dosing, and increase bioavailability. Transdermal delivery is an attractive alternative delivery technique and has a variety of advantages compared to oral, intravenous delivery strategies. Polymeric nanofiber is an excellent choice for transdermal drug delivery systems and it act as drug depots. Polymeric nanofibers were fabricated from electrospinning, which forms interconnected pores in nano-range, high surface-area-to-volume, low hindrance, flexible, and higher mechanical strength. In Figure. 1.7, breast adipose tissue surrounded by cancer, it is very difficult for drugs to reach the specific site. Polymeric nanofibers enhance the nanomaterial skin's internalization ability and it permeates through stratum corneum pores near hair follicles. So, the polymeric nanofiber is a safe, non-invasive, effective *in-situ* delivery technique for transdermal therapeutic application [47,48].

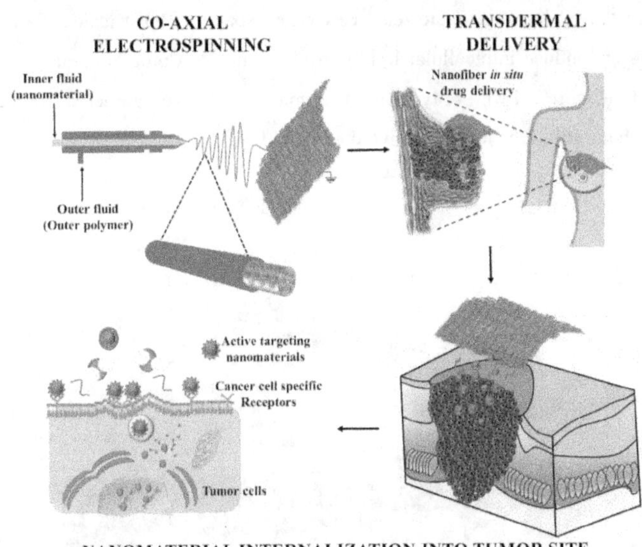

Figure.1.7. Transdermal drug delivery from polymeric nanofiber for breast cancer.

1.5.5. Alternative therapy

It is necessary to develop alternative treatment or combinational treatment strategies due to the failure of conventional drugs or regimens. The strategy should combine with already used drugs/antibiotics with the utilization of alternative mechanisms of action. For instance, modulating the immune response against the disease along with the drug action. Polymeric nanomaterials have an excellent role to bring novel treatment strategies into the intended position in the medical field with excellent efficiency. Polymeric nanomaterial can modulate the immune response, by delivering antigens to dendritic cells which possess a vital role in initiating immune responses. PCPP polymers are particularly adept at enabling immunotherapeutic approaches for the reason they can be engineered to have different properties, surface ligands, and encapsulated agents. Moreover, the attractive administrable routes are easily taken up by dendritic cells and have shown promising immunotherapy [49,50].

A combination of the most common phototherapy methods-PTT and PDT with chemo drugs will improve treatment outcomes and adherence. In PTT, a photoactive agent is triggered by both specific band light and heat release to selectively target the tissues/cells. In PDT, photosensitizers which are activated by light, generate reactive oxygen species (ROS) inside the cells. The initiated oxidative stress can induce intracellular lipid peroxidation, DNA/protein damage, and finally leading to cell death (Figure. 1.8). Polymeric nanomaterials have garnered higher attention for providing promising solutions for alternative therapy [51,52].

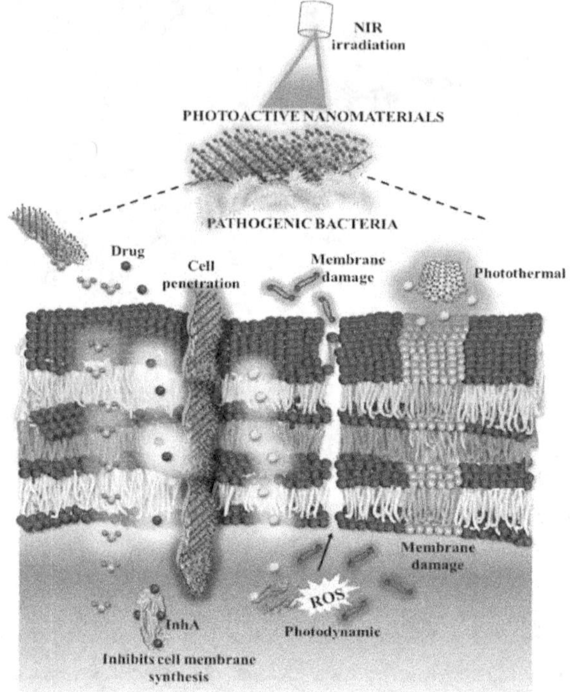

Figure.1.8. Alternative/combinational therapeutic strategy.

1.5.6. Polymeric scaffold

Porous polymeric scaffolds play an important role in BTE and regenerative medicine. Poly(organo)phosphazenes proven suitable for bone regeneration due to the porous three-dimensional (3D) structure. Biodegradable scaffolds can support bone growth and mimic the natural extracellular matrix (ECM). It is also bio-erodible material and has mechanical properties like natural tissues. To increase mechanical strength further, PCPP polymer was reinforced with ceramic material [53-55]. Especially, ceramic material from natural origin has gained attention due to its adequate sources of biomolecules, rare bioactive compounds, and bioceramics, etc. Marine sponges, eggshells, crab shells, corals, snail shells, shrimp shells, fish-bone are available abundant in nature. It composed of organic, inorganic compounds, and have been proven to improve cellular functions, such as adhesion, growth, and secretion. It can be able to produce a bio-silica structure with a more amount of trace elements, such as S, Al, K and Ca naturally. Ceramics from biological origin also have properties like fine-scale microstructure, alterable crystalline property, porosity, morphology, and mechanical property [56].

1.5.7. Polymer as coating material

Titanium (Ti) was the most commonly used bone substitute in the medical field. It has outstanding mechanical strength, bioinert, highly stable in physiological condition, sterilizable, and cheap. Although it is a biocompatible material, the poor integration ability with surrounding tissue is the major drawback of titanium implants [57]. Coating of Ti implant with polymer is commonly used to overcome the issue and electrophoretic deposition (EPD) was an excellent choice for coating biomaterials. Polymer facilitates osteointegration and controls the rejection process. Further to reduce the bone mass loss from the implant, the polymer was combined with natural ceramic. It improves the life span of the implant, the cells identify binding points in the metal implant, biomineralization improves cell adhesion, and growth (Figure. 1.9). The polymeric nanocomposites also help in carrying antibacterial agents to prevent the formation of biofilm over the metal implant [58,59].

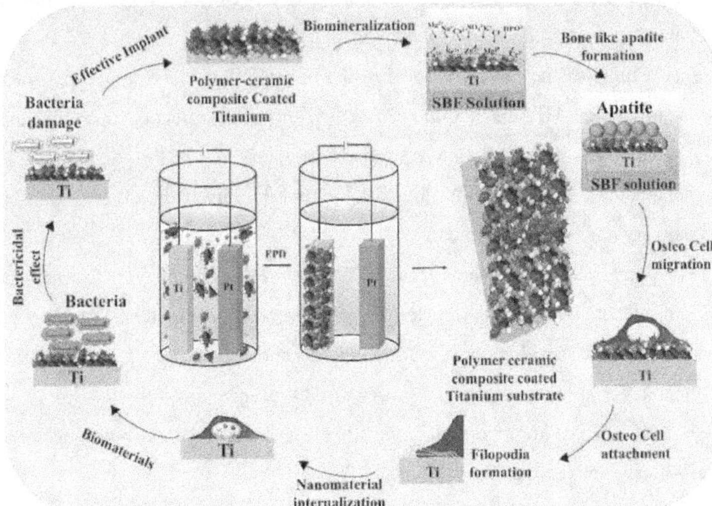

Figure. 1.9. Effective application of polymer coating of Titanium substrate via EPD.

1.6. Objective of the thesis

A versatile multifunctional nanomaterial that inherently possesses a wide variety of therapeutic applications is highly desirable for advanced studies. Over the last few decades, polymers have given rise to significant breakthroughs towards the employment of nanomaterials in several biomedical applications, like cancer therapy, infectious disease, tissue engineering, sensors, and bio-imaging. Among the several polymers, poly(organo)phosphazenes shows some interesting properties like biocompatibility, tunable mechanical properties, polydispersity, structural variation, high drug loading ability, controlled drug release, stability, specificity, inert, and less side effect. Polymer bearing with either positive or negative charges have displayed an immunostimulatory activity. A key purpose of the present work is to develop a multifunctional PCPP polymeric nanomaterial for variety of therapeutic applications like cancer, TB, and bone-tissue engineering. To achieve the desired property, PCPP was combined with natural/synthetic polymer, inorganic, liposomes, carbon-based materials, and biomaterial. The addition of these components can enable specific functionalities like higher drug loading ability, pH/thermoresponsive property, sustained drug release, targeted drug delivery, implant coating, and alternative therapeutic modality.

The following works with specific objectives on developing multifunctional poly(organo)phosphazenespolymeric nanomaterial for cancer, TB, and bone-tissue engineering application were discussed detailly in each chapter of the thesis

1. To evaluate the dual drug loading ability of poly(organo)phosphazenes multilayer nanoparticles (MLNPs) and *in vitro/in vivo* anticancer effect against oral cancer.

2. Development of pH/thermo responsive property on PCPP-polylactic acid hybrid polymer nanomaterial by fine controlling the hydrophilic and hydrophobic chains for enhanced drug delivery against breast cancer.

3. Unveiling cholate functionalized PCPP polymeric nanomaterial for nuclear-targeted delivery and selectively induction of cytotoxicity in breast cancer cells.

4. Design and development of mucilage nanofiber patch loaded active targeting PCPP nanomaterial for transdermal delivery of PTX drug against breast cancer.

5. Fabrication of immunomodulating PCPP-arginine layered liposome vehicle for RIF and isoniazid delivery against pulmonary TB.

6. Combination of photoactive nanoagent-rGO-MoS$_2$ (reduced graphene oxide- Molybdenum disulfide), ICG, with PCPP polymer for effective photothermal/photodynamic and chemotherapy against TB.

7. Assembly PCPP polymeric scaffold via salt-leaching technique and dip-coated with a variety of natural snail shell HAP for osteoinduction.

8. EPD of halloysite nanotubes-PCPP polymer on Ti-6Al-4V screw for evaluating the mechanical, adhesion strength, and regeneration ability of defective bone.

In this chapter, the polymer preparation, various forms of polymeric nanomaterial fabrication methodology, and characterization techniques involved with nanomaterial development are elaborated. Poly(organo)phosphazenes polymers were synthesized by thermal polymerization of hexachlorocyclotriphosphate. This reaction forms poly(dichlorophosphazene), which reacts with propyl-p- hydroxybenzoate to form poly(bis(carboxyphenoxy)phosphazene) polymers. The polymeric nanomaterial and drug-loaded polymeric nanomaterial were fabricated by an emulsion solvent evaporation method. Fabricated polymeric nanomaterials (with and without drugs) were subjected to preliminary characterizations to evaluate their physicochemical properties. Investigations on *in vitro* behavior of polymeric nanomaterials like swelling, weight loss, hydrolytic behavior were analyzed. Polymeric nanomaterial interaction with different types of hydrophobic/hydrophilic drugs was studied by drug loading and encapsulation studies. Polymeric nanomaterial effective drug release ability was performed by dialysis method. *In vitro* anticancer drug release ability of polymeric nanomaterial was evaluated by cancer cells like MCF-7, KB-cells. The major techniques like MTT assay, AO/EB staining, ROS, and MMP assay were performed. The alteration in the protein level after treatment with polymeric nanomaterial was examined by western blotting. The antibacterial effect of polymeric nanomaterial was investigated by the disc-diffusion method, MIC, and MBC measurement. The effective bone regeneration capability of polymeric nanocomposites was studied by biocompatibility, cell proliferation, clonogenic assay, and cell adhesion studies.

2.1. Synthesis of poly[di(sodium carboxyphenoxy)phosphazene]

Firstly, 0.1 g of hexachlorocyclotriphosphate dissolved in dichloromethane (DCM). Followed by the addition of chlorophosphoranimine to the mixture, it was stirred at room temperature for 4 h. Subsequently, DCM was then removed under reduced pressure to form poly (dichlorophosphazene). Secondly, the solution of propyl-p- hydroxybenzoate (0.3 M) in diglyme was heated at 110 °C for 30 min with the presence of nitrogen. To this, a suspension of sodium hydride (12.2 g) was slowly added to the propyl-p-hydroxybenzoate solution and stirred under nitrogen to form the sodium salt of propyl-p- hydroxybenzoate [60]. Thirdly, poly(dichlorophosphazene) (1.12 g) was slowly added to the above solution and temperature annealed at 120 °C for 10 h and reaction temperature was reduced to 85 °C and 300 mL of potassium hydroxide was added and allowed to 2 h. The synthesized poly[di(sodium carboxyphenoxy)phosphazene] (PDCPP) polymer was further precipitated by adding 600 mL of sodium chloride (30 %) solution [61]. The molecular weight and polydispersity index (PDI) of the synthesized polymer was measured by gel permeation chromatography (RID 20A, Shimadzu, Japan).

2.2. Synthesis of poly(bis (carboxyphenoxy) phosphazene)

Poly(bis(carboxyphenoxy)phosphazene) (PCPP) was synthesized by thermal bulk polymerization of hexachlorocyclotriphosphate [62]. Briefly, 0.1 g of hexachlorocyclotriphosphate was dissolved in 25 mL of DCM for 15 min. Frequently, chlorophosphoranimine was added and the mixture was stirred at room temperature for 4 h. DCM was then removed under reduced pressure to give poly (dichlorophosphazene). A polymer of 4.0 g dissolved in 400 mL of THF. Propyl-p-hydroxybenzoate of 10 g was dissolved in 100 mL THF and then added to the poly (dichlorophosphazene). Cesium carbonate of 16.9 g was then immediately added to the reaction mixture. The reaction proceeds at room temperature for 3 days. Further, the solution was concentrated and precipitated into water 3 times and hexane once. The solvent was removed under reduced pressure to get a solid polymer.

2.3. Preparation of drug-loaded polymeric nanomaterial

Hydrophobic drug (10, 20, 30, 40, 50 mg) was dissolved in 5 mL chloroform stirred overnight at room temperature. For the water phase – 25, 50, 75, 100 mg of the polymer was dissolved in 10 mL water for 1 h. The two solutions were mixed with a probe-sonicator (Sonics, Vibracell) for 5 min. Subsequently, the emulsion was put in a flask and kept stirring, allowing complete evaporation of chloroform for 24 h. Once the evaporation step was completed, nanomaterials were washed twice with water and collected via centrifugation at 8000 rpm for 20 min. Subsequently, they were freeze-dried for 2 days and stored at 4 °C [63].

2.4. Characterization techniques
2.4.1. Nuclear Magnetic Resonance

The ^{1}H, ^{13}C high-resolution Nuclear magnetic resonance (NMR) is an important study of higher-order stereochemical placement of polymers. Also, carbon is the most essential atom constituting the skeletal bonds of the polymer chain. ^{1}H, ^{13}C NMR measurements were observed in a Bruker (Avance) NMR spectrometer (400 MHz proton frequency) at 25 pulses 40° with CDCl$_3$, DMSO-d6 as solvents, and tetramethylsilane (TMS) as an internal reference.

2.4.2. Fourier Transform-Infrared Spectroscopy

Fourier Transform-Infrared Spectroscopy (FTIR) was used to determine the specific functional groups of nanomaterial samples by measuring the absorption of infrared radiation. Each sample was subjected to mix with potassium bromide (KBr) in the ratio of 1:100 (Sample: KBr) to form a pellet using a load of 10 tons/cm^2. Each sample measured on wavenumber 400 to 4000 cm^{-1} and 1000 scans were collected for each sample at a resolution of 4 cm^{-1}. The spectrum was performed using the Perkin Elmer GX1 FT-IR spectrophotometer.

2.4.3. X-ray Diffraction

X-ray Diffraction (XRD) is an analytical technique used to identify the grain size, composition of the material, crystalline nature, etc. Further, it also provides information on lattice orientation as well as the unit cell dimensions of the crystalline materials. Here, XRD was used to determine the crystalline phase of PCPP, PCPP-conjugates, nanocomposites samples. During analysis, the samples were finely grounded into powder using mortar and pestle. XRD patterns for every sample were captured on a powder diffractometer (XRD, Bruker AXS D8) with a nickel-filtered Cu Kα X-ray beam at 40 kV and 25 mA ($\lambda = 0.15418$ nm). Every sample was scanned in the 2θ range of 10° to 80°.

2.4.4. Dynamic Light Scattering and Zeta potential measurement

Dynamic Light Scattering (DLS) is a technique used to measure the hydrodynamic size of the nanomaterials. It also measures the surface charges (zeta potential) and stability using an electrophoretic light scattering spectrophotometer. For the particle size analysis, nanomaterial samples were diluted 100-fold with solvents before analysis, and samples were subjected to light scattering at 90° angle. The hydrodynamic size of samples was measured by a (Delsa TM Nano, Malvern, UK). Similarly, the zeta potential measured using electrophoretic cells, and values were calculated from the mean electrophoretic mobility.

2.4.5. Thermal Gravimetric analysis

Thermal Gravimetric Analysis (TGA) was executed to determine the mass difference and phase transition of a nanomaterial at varying temperatures. TGA was performed using Mettler-Toledo instrument, Greifensee, Switzerland. Samples of 15-20 mg were placed in the pan with a heating rate of 10 °C/min from 25 °C to 700 °C under a nitrogen atmosphere.

2.4.6. Scanning Electron Microscopy with energy-dispersive X-ray analysis

Scanning Electron Microscopy (SEM) is a mostly used electron microscope to evaluate the size, surface topography, and morphology. Nanomaterial sample morphology was performed using SEM (HITACHI SU-6600) and elemental composition was observed with energy dispersive X-ray analysis (EDAX, Horiba 8121-H). The sample was diluted 100-fold with water and sonicated for 20 min. The dispersion was cast on a glass slide, allowing for drying. Then, dried samples were sputter-coated with gold and visualized through a computer.

2.4.7. Transmission Electron Microscope

Transmission Electron Microscope (TEM) analysis is used to visualize the nanomaterial in higher magnification, nano-sized material, and topographic details on the surface of the materials. TEM analysis was performed in JEM-100, JEOL. The samples were diluted to 1:10000 ratios with water and sonicated for 20 min. The well-dispersed samples were added to the copper grid. Then, samples were subjected to the accelerating voltage of 200-220kV in various magnifications.

2.4.8. Atomic Force Microscope

The surface topography of nanomaterials was analyzed using Atomic Force Microscope (AFM). Tapping mode AFM phase image was collected to examine the surface morphology of the samples. A drop of a diluted sample of 0.1 mg/mL was coated on a glass substrate and kept for drying. To get a high-resolution image, a silicon cantilever with a length of 125 µm, a resonance frequency of 270 kHz (Digital Instrument S3000 AFM). The cantilever oscillates at slightly below its resonance frequency with amplitude from 20 to 100 nm. The tip slightly taps on the sample surface during scanning.

2.4.9. Compressive strength

The compressive mechanical studies were performed using the Sansi CMT6503 (Shenzhen, China) universal testing machine of load cell- 5 kN. The stress was carried at the 10 % strain, crosshead speed of 1.0 mm min^{-1}. The samples of 10 × 10 mm were examined three times.

2.5. Physiochemical characterization

2.5.1. Buffer preparation

MES Buffer (2-ethanesulfonic acid, 0.1 M) preparation, dissolve 19.2 g of MES free acid in 900 mL of water and adjust the pH (5.2-6.0), makeup volume to 1,000 mL. Phosphate buffer saline (PBS) of 1000 mL was prepared by dissolving, NaCl (8.0 g), KCl (200 mg), Na_2HPO_4 (1.44 g), KH_2PO_4 (245 mg) in 800 mL water and adjust the pH 7.4. Sodium acetate buffer prepared by a mixture of 0.2 M $CH_3COONa.3H_2O$ and acetic acid (0.1 M), and adjust the solution to desired pH (4.0, 5.0, 6.0). The sodium carbonate buffer was prepared by a mixture of 0.1 M $Na_2CO_3.10\ H_2O$ and $NaHCO_3$, adjusting the pH 9.0. SBF was prepared by dissolving, NaCl (8.035 g), $NaHCO_3$ (0.355 g), KCl (0.225 g), $K_2HPO_4.3H_2O$ (0.231 g), $MgCl_2.6H_2O$ (0.311 g), $CaCl_2$ (0.292 g), and Na_2SO_4 (0.072 g) in distilled water (1000 mL). The solution was buffered at pH 7.4 with Tris (hydroxymethyl aminomethane) $((CH_2OH)_3\ CNH_2)$ (6.118 g) and 1 M hydrochloric acid (HCl) (39 mL). A lysis buffer with KH_2PO_4 (20 mM, pH 7.4), NaCl (150 mM), and imidazole (10 mM) was used. Krebs buffer was prepared by NaCl (1.26 M), KCl (25 mM), $NaHCO_3$ (250 mM), NaH_2PO_4 (12 mM), $MgCl_2$ (12 mM), $CaCl_2$ (25 mM) and adjust to pH 7.2.

2.5.2. Swelling studies

To determine the response of polymeric nanomaterial *in vivo* conditions, swelling, weight loss, and hydrolytic degradation studies were performed. For swelling studies, PCPP polymeric nanomaterial of 0.05 g was immersed in sodium acetate solution (pH 4.0), phosphate buffer solution (pH 7.4), and sodium carbonate solution (pH 9.0) at temperature 37 °C until a swelling equilibrium was attained. The weight of swollen samples was measured against time after the excess surface water was removed by gently tapping the surface with a dry piece of filter paper. The swelling ratio (%) was calculated according to the following equation [64].

$$Swelling\ ratio\ \% = \frac{(Weight\ of\ wet\ nanomaterials - Weight\ of\ dry\ nanomaterials)}{(Weight\ of\ dry\ nanomaterials)} \times 100$$

2.5.3. Weight loss

The nanomaterial weight loss was examined by immersing the dry nanomaterial of weight in a PBS buffer. In predetermined time intervals, the samples were removed from the buffer, rinsed in distilled water, and weighed [65]. The weight loss (%) was calculated by the following equation,

$$Weight\ loss\ (\%) = \frac{Weight\ after\ immersion - Initial\ dry\ weight}{Initial\ dry\ weight} \times 100$$

2.5.4. Biomineralization test

The polymeric scaffold/implant ability on biomineralization was investigated using SBF, which mimics the ion concentrations in blood plasma. Scaffolds were immersed in the SBF solution for different time intervals and maintained at 37 °C. The SBF was changed each day with fresh buffer and after incubation, the samples were removed, rinsed in water, and dried for SEM and elemental mapping studies are taken from the same area at different defocii.

2.6. Polymeric nanomaterials-drug interaction studies
2.6.1. Evaluation of drug encapsulation efficiency

The encapsulation efficiency (EE) of a drug was calculated by UV-vis spectrophotometry at a specific λ-max value of each drug represented in Table.2.1. respectively. The polymeric nanomaterial of 10 mg was dissolved in the 10 mL PBS under vigorous vortexing [66]. The solution obtained was filtered and the concentration of drugs was measured using an UV-vis spectrophotometer (UV1800, Shimadzu, Japan). The EE % was calculated as follows:

$$\% \, EE = \frac{(Total \; weight \; of \; nanomaterial - Free \; nanomaterial \; weight)}{(Total \; weight \; of \; nanomaterial)} \times 100$$

Table.2.1. Specific λ-max value of different drug.

Drug	λ-max value (nm)
Cisplatin	310
Chrysin	348
PTX	229
RIF	337
IZN	263
ICG	780

2.6.2. Drug release studies

The *in vitro* drug releases from the nanomaterials were evaluated by the dialysis method. Briefly, nanomaterial of 10 mg placed in a dialysis membrane and placed in of 10 mL release medium of different pH. The release reservoir was kept at constant stirring. At predetermined time intervals, 2 mL of buffer solution was taken for investigation. Simultaneously, 2 mL of fresh buffer solution was replaced to maintain the solution quantity and identical pH value [67]. The cumulative drug release from the nanomaterial was measured by the respective λ-max value of the drug by UV-vis spectrophotometer.

2.7. *In vitro* anticancer studies
2.7.1. Cell culture

KB cells (Human oral cancer cell lines), MCF-7 cells (Human breast cancer cell line) were procured from National Centre for Cell Science, Pune, India. Dulbecco's modified Eagle's medium was supplemented with 10 % fetal bovine serum and 1 % penicillin-streptomycin solution used as the cell culture medium. Cells were grown in a humidified environment at 37 °C with 5 % CO_2.

2.7.2. Cellular uptake

The study was performed to determine the nanomaterial ability on the target cell penetration. The FITC loaded polymeric nanomaterial was used to visualize the cellular uptake efficiency. The 4×10^3 cells per well were seeded in a six-well plate and incubated at 37 °C along with 5 % CO_2 for 24 h. After incubation, the medium was replaced, and added the medium containing FITC loaded polymeric nanomaterial of 25 µg/mL. The treated cells were incubated at 37 °C along with 5 % CO_2 for different time intervals [68]. Finally, cells were washed with PBS to remove unbounded particles on the surface, the cells were observed under fluorescence microscopy (TCF100, Nikon).

2.7.3. Cytotoxicity assay of cancer cells

The cancer cell death was measured by MTT (3-(4, 5-Dimethylthiazol-2-yl)-2, 5- Diphenyltetrazolium bromide) and IC_{50} concentration denotes the half-maximal inhibitory concentration. Cancer cells were seeded in a 96 well plate at a density of 4×10^3 cells per well were incubated for 24 h. After incubation, media containing polymeric nanomaterial at various concentrations were added. Cells were maintained at 37 °C along with 5 % CO_2 for the respective period. Then, cells were washed with PBS buffer, MTT (5 mg/mL) added to each well, and kept at 37 °C for 3 h. Further, DMSO (Dimethyl sulfoxide) of 0.5 mL was added and optical density was measured using a spectrophotometric plate reader (2030 multilabel reader Victor X3, Perklin Elmer) at 595 nm [69].

$$Cell\ viability\ (\%) = \frac{OD\ of\ treated\ sample}{OD\ of\ control\ sample} \times 100$$

2.7.4. AO/EB staining

Double staining dye (AO/EB) was used to differentiate normal and apoptotic cells. Cancer cells were treated with different concentrations of polymeric nanomaterials for 24 h. Then, it was washed with cold PBS and stained with a mixture of AO (100 µg/mL) and EtBr (100 µg/mL) at room temperature for 10 min. The stained cells were washed twice with PBS and observed under a fluorescence microscope.

2.7.5. Hoechst Staining

Hoechst 33342 was a nuclear stain used to detect chromatin condensation. After cancer cells treatment with polymeric nanomaterial for 24 h, cells were harvested and washed with PBS. Treated and untreated cells are exposed to Hoechst 33342 (10 µg/mL) at room temperature in the dark for 20 min. After washing, the cells with PBS were observed under a fluorescence microscope.

2.7.6. Determination of Reactive Oxygen Species

To determine the ROS levels, polymeric nanomaterial was treated with cancer cells and incubated for 24 h. The treated and untreated cells were loaded with 100 µM DCFH-DA (Dichloro-dihydro-fluorescein diacetate) for 1 h. Cells were then washed with PBS. Fluorescence images were recorded under a fluorescent microscope.

2.7.7. Measurement of Mitochondrial Membrane Potential

Cancer cells were treated with polymeric nanomaterial and incubated for 24 h. Treated cells were washed twice with PBS, and the cell pellet was then resuspended in 2 mL of fresh incubation medium containing 2.0 µM rhodamine. The cells were incubated at 37 °C for 20 min with gentle shaking. Cells were collected by centrifugation and washed twice with PBS, then analyzed by fluorescence microscope. Untreated cells were maintained as control. The uptake of the cationic fluorescent dye rhodamine-123 (Rh-123) has been used for the estimation of mitochondrial membrane potential (MMP).

2.7.8. Western blotting

Cancer cells after treatment with nanomaterial of IC_{50} concentration were lysed with lysis buffer and centrifuged at 12,000 rpm for 10 min at 4 °C. Supernatant protein was resolved in SDS-PAGE gels and transferred to a PVDF transfer membrane (Bio-Rad) at 100 V and 350 mA for 1 h. After treatment with primary antibodies like rabbit anti-bcl-2 antibody (1:500), rabbit anti-bax antibody (1:500), rabbit anti-caspase 3 antibody (1:500) or rabbit anti-β-actin antibody (1:500), the membrane was subsequently washed with TBST (1 X TBS, 0.1 % Tween-20) three times for 10 min. The membrane was then incubated with secondary mouse anti-goat peroxidase-conjugated antibodies (Cell signaling technologies, USA) for 1 h at 37 °C. After washing the membrane with TBST, they were developed by the diaminobenzidine chromogenic detection method and then scanned [70].

2.8. *In vitro* antibacterial studies
2.8.1. Cell culture

In-vitro biocompatible test conducted using U937 cells (Human macrophage cells) acquired from the National Centre for Cell Science (NCCS), Pune, India. RPMI 1640 containing 10 % FBS, 2 mM glutamine, and 1 % Fungi-bact were used as the cell culture medium. Cells were grown in a humidified environment at 37 °C with 5 % CO_2.

2.8.2. Bacterial culture

Gram-positive-*Bacillus subtilis* (MCC 2511), *Staphylococcus aureus* (MCC 2408), *and Clostridium sp.* (MTCC 1349), Gram-negative- *E.coli* (MCC 2413), *Klebsiella pneumoniae* (MCC 2451), *Salmonella typhi* (MTCC 3224) bacteria were acquired from the Microbial Type Culture Collection and Gene Bank (MTCC). Bacterial culture was inoculated into LB broth. And, the *Mycobacterium tuberculosis* (ATCC 25618) were acquired from the American Type Culture Collection (ATCC). Bacterial sample was grown in the Middlebrook 7H9 broth.

2.8.3. Disc-diffusion studies

Antibacterial activity of polymeric nanomaterial was tested against both Gram-positive and Gram-negative bacteria using the disc diffusion method. Bacteria cultures with a cell concentration of approximately 10^9 cells/mL were diluted 10 times. Then, 200 µL of diluted culture was spread on an agar plate, and then 6 mm disks contain polymeric nanomaterials punched gently to the plate with sufficient distance. The plates were incubated at 37 °C for 24 h; after incubation time, the zone of inhibition diameter was measured, and experiments were performed as the mean of triplicates [71].

2.8.4. Antimycobacterial effect of nanomaterials

The polymeric nanomaterial antimycobacterial effect against the *Mycobacterium tuberculosis* was evaluated. Different concentrations of polymeric nanomaterial added to the bacterial culture. Then, growth-inhibitory was measured by optical densities at 600 nm [72]. The percentage (%) of the growth inhibition was determined by the following equation,

$$Growth\ inhibition\ (\%) = \frac{(B_a - B_b)}{B_a} \times 10$$

Here, B_a was an average of six replicates of light absorption values at 600 nm wavelength of the controls, and B_b was an average of six replicates of light absorption values at 600 nm wavelength of the treated samples.

2.9. *In vitro* bone regenerative studies

2.9.1. Cell culture

The MG-63 cells (Human osteosarcoma cell line) were procured from National Centre for Cell Science, Pune, India. Eagle's minimum essential medium supplemented with 10 % fetal bovine serum and 1 % penicillin-streptomycin solution used as the cell culture medium. Cells were grown in a humidified environment at 37 °C with 5 % CO_2.

2.9.2. Cell proliferation

To identify the cell attachment, morphology, and proliferation rate of MG-63 cells after treatment with polymeric nanocomposites were visualized by phase-contrast microscope. MG-63 cells of 4×10^4 cells/µL were seeded on the plate and polymeric nanocomposites dispersed in the medium were added. Then, cells were maintained at 37 °C, 5 % CO_2 for a respective period. Finally, polymeric nanocomposites treated cells were visualized by phase-contrast microscope [73].

2.9.3. Clonogenic assay

The clonogenic ability of MG-63 cells after treatment with polymeric nanocomposites was determined by a clonogenic assay. MG-63 cells of 4×10^4 cells/µL were seeded in plate and treated with polymeric scaffold and implant. The culture medium was replaced twice a week. On days 7, 14, and 21 cells were washed with PBS, fixed using methanol, and stained with Crystal Violet. The colonies of cells containing more than 50 cells were classified as a colony originating from one clonal cell.

2.9.4. Flow cytometry

The polymeric nanocomposites biocompatibility was measured by FACS flow cytometer (Beckmann Coulter-Moflo XDP). The samples of 60 µg/mL concentration treated MG-63 cells and maintained at ambient condition. The polymeric nanocomposites treated cells were mixed with PI stain. Gating of live and dead cell population as executed on SYTO 9/propidium iodide scatter plot, and event counts are given per cm^2.

2.9.5. Osteoblast cells adhesion on TSS-HAP/PCPP3/PCL scaffold

MG-63 cells adhesion, structural morphology after treatment with polymeric nanocomposites were evaluated by SEM. Cells were seeded on the polymeric nanocomposites containing medium and maintained at 5 % CO_2, 37 °C for 5 days. Polymeric nanocomposites treated cells were fixed with glutaraldehyde of 2.5 % and washed with PBS. Dehydration with ethanol of 30, 70, 90, 96 and 100 % (v/v) concentrations for 10 min. After complete drying of samples, it was visualized under SEM imaging.

2.9.6. Statistical analysis

All the measurements were presented as mean ± standard deviation (SD). Three independent experiments are performed with three replicates for each sample to prove reproducibility. The statistical analyses of differences analysed by independent t-test and post hoc test for one-way ANOVA. A value of $p < 0.05$ was considered to be statistically significant.

3.1. Introduction

Oral cancer therapy using nanocarriers provides an opportunity to enhanced the permeability and retention (EPR) effect; accumulate the drug at targeted sites, sustain blood circulation and reduce off-target delivery [74,75]. Despite these desirable features, certain drawbacks such as uncontrolled release, poor cellular uptake, inefficient tumor penetration, and lack of perinuclear drug release need to be addressed. Initially, nanoparticles fabricate for functions like cross barriers to deliver the loaded drug at target sites [76-78]. For instance, MLNPs can maximize efficacy by the multi-layered protection of drug, optimum size, triggered drug release, and high cell penetration ability. Several strategies like seeded emulsion, dispersion polymerization, hetero-coagulation, and layer-by-layer deposition are available for the development of multilayer structures [79-81]. In this work, MLNPs were assembled via electrostatic interactions of anionic/cationic PELs. It can able to fabricate with controlled properties like thickness, composition, roughness, high drug loading, porosity offering biomedical and pharmaceutics applications. The high amount of drug-loaded MLNPs response to the acidic stimuli, which could increase the release of drugs at the targeted site [82,83]. Amphoteric MLNPs are eventually protonated at the acidic pH, which destabilizes the multilayer resulting in the release of the therapeutics at the targeted site. It is paramount fact to be considered for an ideal drug delivery system with polymer superior protection and minimize premature drug release using covalent links [84,85].

3.2. Previous work done

At present most of MLNPs are dominated by polymers like Poly (diallyl-dimethyl-ammonium chloride), poly (ethyleneimine), poly (allylamine) hydrochloride, poly(styrenesulfonate), poly (vinylsulfate) and poly (acrylic acid), natural PELs like alginate, chitosan are used for layer-by-layer deposition techniques [86,87]. On the other hand, the responsiveness of MLNPs with a desirable size range was achieved by ionic strength, pH of the solution, PELs concentration [88]. However, MLNPs was encapsulated with the combinational drug against oral cancer were rarely reported up to now. This modular design enables the development of a core encapsulated with multiple drugs.

3.3. Novelty

In this, Cisplatin and Chrysin were incorporated in the porous mineralized $CaCO_3$ nanoparticles (CCNPs) core and polymeric shell. Calcium carbonate is the most abundant inorganic particle, which has been well utilized as a template core for the preparation of multi-layer nanoparticles. Compared with other porous materials, $CaCO_3$ is cost-effective with no addition of polymer additive, enabling it to be embedded into functional molecules [89,90]. CCNPs were able to control the release of anticancer drugs against oral cancer. Bee-related natural products such as honey and wax have high therapeutic applications. Chrysin (5, 7- dihydroxyflavone) is one of the natural bioactive dietary flavones abundantly found in bee wax and it has medicinal properties such as anticancer, anti-inflammatory, antioxidant, hepatoprotective, antimicrobial, and anti-diabetic effects. More evidence suggests that the combination of the natural compound with anticancer drug delivery via nanocarriers shows synergistic drug action and a promising therapeutical approach for oral cancer [91].

3.4. Objective

The present study describes the detailed mechanistic aspects of the assembly of MLNPs with optimum size, high drug loading, and controlled drug release ability. And also, examine the physicochemical characterization of MLNPs using *in vitro* and *in vivo* models of oral cancer. To achieve this, MLNPs were fabricated with PDCPP - poly (diallyl dimethyl ammonium chloride) (PDADMAC) alternative layer on CCNPs surface. Using this depiction, chrysin/cisplatin was encapsulated in the core-shell layers, to form a multi-drug delivery system. The polymeric electrostatic interaction, stability, optimum size, structure, and topography were studied. In addition, a comparison of single and dual drug-loaded MLNPs in sustained drug release, cell penetration, cytotoxicity was examined in KB cells. Finally, *in vivo* hamster animal model was used to identify the tumor regression and biocompatibility efficiency of MLNPs.

3.5. Experimental techniques
3.5.1. Preparation of drug-loaded CCNPs coated by multilayer polymer

Cisplatin and chrysin-loaded CCNPs were prepared at room temperature with some modification. For encapsulation of drug, cisplatin was dissolved in 0.9 % saline solution and chrysin was dissolved in water: ethanol mixture of 1:0.5 ratios. This dissolved drug solution of 5 mg.mL^{-1} was mixed with 0.01 M CaCl$_2$ solution and the resulting mixture was added dropwise to 0.01 M Na$_2$CO$_3$ solution and vigorous mixing at different (800, 1400, and 2000) rpm for 12 and 24 h. After centrifugation, drug-loaded CCNPs were suspended in 0.5 M NaCl solution and were then added to PDADMAC (1 mM) solution for 30 min. The excess polyelectrolyte solution was removed by two times centrifugation (12000 rpm × 15 min) and washing with deionized water. Subsequently, CCNPs coated by a monolayer of polymers were re-suspended in PDCPP for 30 min, followed by two times centrifugation and washing.

3.5.2. FITC loaded MLNPs

The cellular uptake of the MLNPs was studied using fluorescence microscopic investigation of FITC loaded MLNPs on using KB cells. Briefly, 30 mg of FITC was added to $CaCl_2$ solution for 15 min and the mixture was added to Na_2CO_3 solution mixed constantly for 12 h. FITC loaded CCNPs were covered with PDADMAC and PDCPP repeating alternative layer to form MLNPs.

3.5.3. Animals studies

Male golden Syrian hamsters aged 8 to 10 weeks were obtained from the National Institute of Nutrition, Hyderabad, and maintained in grilled cages individually under stress-free environment (RT=22±2 °C, light/dark cycle 12 h) with a standard pellet diet. All the animal experimentations were conducted after getting approval from the Institutional Animal Ethics Committee (AIEC) of Cancer Institute (WIA) (Ref No. 1-WIA20210903 Dt.18.09.2021).

3.5.4. *In vivo* antitumor effect of MLNPs

The animals were divided into seven groups with six hamsters each group as follows. Group 1-control hamsters were painted with liquid paraffin alone on their left buccal pouches. Animals have treated with 0.5 % solution of 7,12-Dimethylbenz[a] anthracene (DMBA) in liquid paraffin alone and tumor sizes reached up to 200-250 mm^3. One group was selected to be cancer group without any treatment (group-2). Tumor-induced hamsters (groups 3, 4, 5, and 6) were orally given MLNPs, cisplatin/chrysin MLNPs, and both drugs loaded MLNPs (at a dose 100 mg/kg body weight) dissolved in 5 % DMSO respectively, thrice a week on alternate days to DMBA application. In group 7, animals were treated with MLNPs alone at 100 mg/kg body weight. The dose of drugs used in this study was chosen based on a dose-response study [92,93]. The response to MLNPs therapy on hamsters over different groups was monitored for tumor incidence, tumor volume, and tumor burden.

3.5.5. Histopathological analysis

The experiment was terminated at the end of the 16th week and animals were sacrificed by cervical dislocation after an overnight fast. The buccal pouches were excised and the tissues were fixed in 10 % formalin. Tissue embedded in paraffin blocks were cut into 3 μm sections and stained with hematoxylin and eosin (H&E) stains and observed under the light microscope (20 X) (Olympus Corporation, Tokyo, Japan).

3.6. Results and discussion
3.6.1. Characterization of PDCPP

A novel PDCPP polymer was successfully prepared by thermal bulk polymerization of poly (dichlorophosphazene) followed by substitution with sodium propyl-p-hdroxybenzoate. The structure of the PDCPP were **C-COONa** (10.12, s), **C-CH** (8.14, d), **C-CH** (7.23, d), **O-CH$_2$** (5.5, t), and CH$_2$-O (72), CH$_2$-O (77), CH=CH (125.10), CH=CH (128.53), CH=CH (133.73), CH=CH (134.73), C-CH$_2$ (141.12), **C-CH$_2$** (147.23), **C-COOH** (173.3) by with ^{1}H and ^{13}C NMR respectively (Figure. 3.1, 3.2).

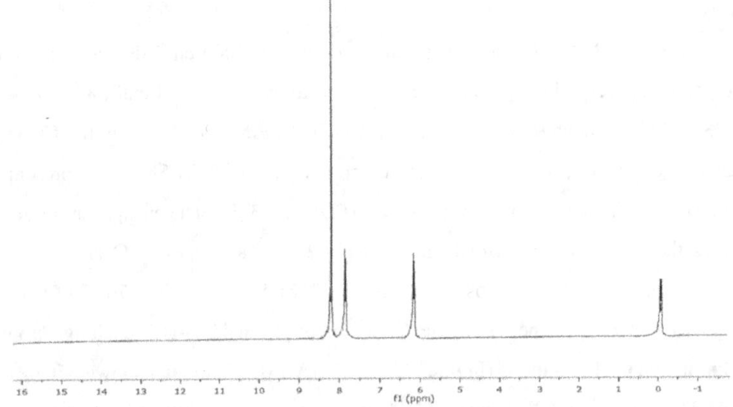

Figure. 3.1. ^{1}H NMR spectrum of Poly (di (sodium carboxyphenoxy) phosphazene) in DMSO with TMS as an internal standard at 300 MHz.

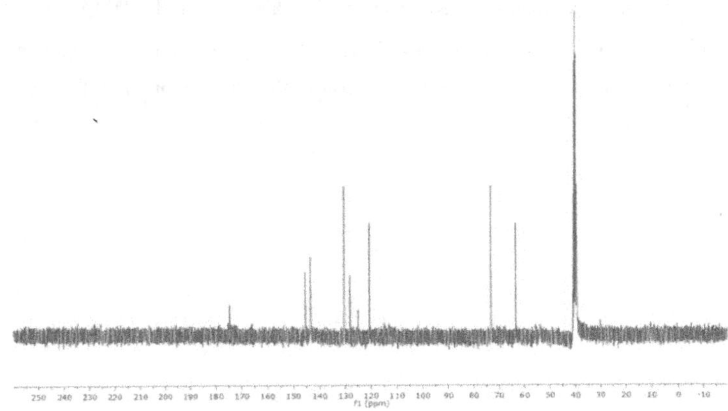

3.6.2. Synthesis and characterization of MLNPs

Drug-loaded negatively charged CCNPs were covalently bound with oppositely charged PDADMAC/PDCPP. The multilayer process was based on the physical adsorption, and strong electrostatic mediated interactions [94,95]. Spontaneous electrostatic adsorption of PDADMAC onto the CCNPs was increased by aqueous NaCl. Further, anionic PDCPP was forming a second layer on the PDADMAC coated CCNPs. To understand the chemical of PDADMAC-PDCPP deposited MLNPs, they were characterized by FTIR spectra.

(i) FTIR analysis

In Figure. 3.3, CCNPs absorption peak that appeared at 3000 cm^{-1} depicting the presence of carbon stretch and band related to carbonate groups were found at 1426.78 cm^{-1}, 875.64 cm^{-1}, 722.49 cm^{-1} (Figure. 3.3. A). The neutral nature of cisplatin facilitates drug loading into the CCNPs with no major change in the characteristic peaks. The absorption peak of 3437.38 cm^{-1} represents the N-H stretch, denotes the cisplatin encapsulated in CCNPs (Figure. 3.3. B). The encapsulation of chrysin on CCNPs denotes the formation of narrowband at 1655.12 cm^{-1} attributes C=O and C=C of chrysin respectively (Figure. 3.3. C). The absorption bands 2121.55, 1988.32, 1766.34, and 968 cm^{-1} correspondings to PDCPP polymer aromatic C-H, carboxylic acid O-H stretching, aromatic C=C bending, and aromatic C-H vibrations (Figure. 3.3. D). PDADMAC polymer shown 3568.21, 2528.01, 2092.24, 1788.21, 1236.29, 1192.44, and 597.04 cm^{-1} corresponds to amide N-H stretch, carboxylic acid, aliphatic amines functional group C-N stretch, and alkenes respectively (Figure. 3.3. E). Both cisplatin and chrysin drugs loaded CCNPs coated with PDCPP and PDADMAC polymer consists of strong absorbance peaks at 1407.44 and 3427.46 cm^{-1}, which represents the -OH stretch of the chrysin and N-H stretch of cisplatin (Figure. 3.3. F). It is confirming the formation of the polyelectrolyte layer around the CCNPs.

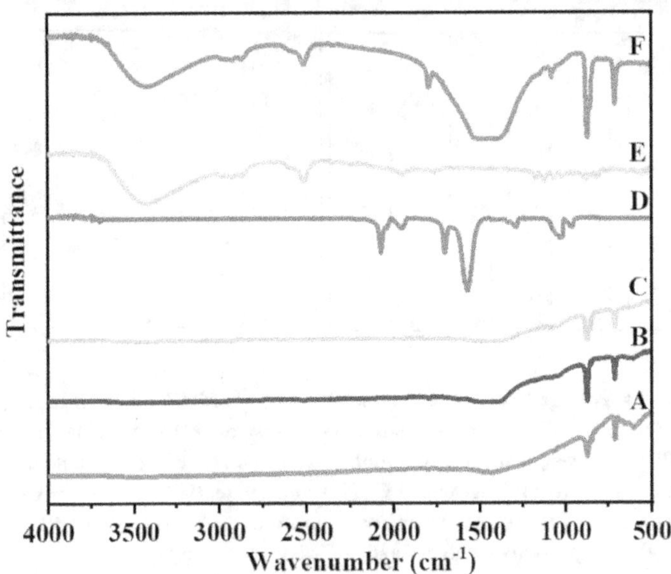

Figure. 3.3. FTIR spectra of (A) CCNPs (B) Cisplatin loaded CCNPs (C) Chrysin loaded CCNPs (D) CCNPs coated with PDCPP layer (E) CCNPs coated with PDADMAC layer (F) Both drugs loaded MLNPs.

(ii) DLS and Zeta potential

PDADMAC-PDCPP deposited MLNPs size and stability were represent in Figure. 3.4. It shows steady increases in the hydrodynamic diameter of MLNPs. In Figure. 3.4. A, the diameter of the CCNPs was 237 nm, the assembly of PDCPP and PDADMAC gradually increases the size of the nanoparticles, but it remained within the optimal range. The addition of two polymers, formed a thin layer on the CCNPs surface denotes strong electrostatic interaction between the PDADMAC and PDCPP. The deposition of PDADMAC (polycation) and PDCPP (polyanion) on the surface of CCNPs were denoted by reversal of the surface charge (Figure. 3.4. B). Initially, CCNPs have -3.6 mV, after the formation of the PDADMAC layer induces a high positive charge. Subsequently, second layer formation showed, the surface charge of particle altered to -5.12 mV, as the PDADMAC and PDCPP layers were successively absorbed.

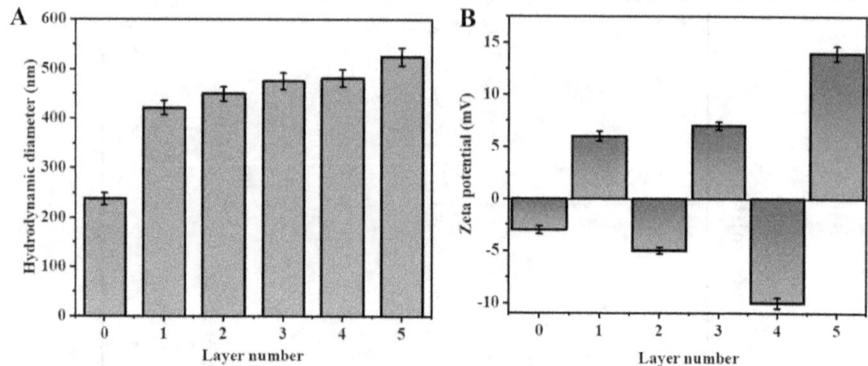

Figure. 3.4. (A) Average hydrodynamic diameters of naked CCNPs as a function of adsorption of polyelectrolyte multilayer. The odd numbers represent - PDADMAC layer and even numbers represent PDCPP layer, respectively (number 0, which represents the naked CCNPs) (B) Zeta potential measurements of drug-loaded CCNPs and sequentially layered PELs CCNPs. The odd numbers represent - PDADMAC layer and even numbers represent PDCPP layer, respectively (number 0, which represents the drug-loaded CCNPs zeta values).

(iii) SEM and TEM analysis

The topographic evidence of CCNPs revealed that each particle was a smooth and spherical shape. The nanoparticles formed in a range of 580, 240, and 120 nm by an effect of (800, 1400, and 2000 rpm) agitation respectively (Figure. 3.5. A, B, and C). The decrease in the size of the nanoparticles and high stability might be due to the agitation speed and time. The drug-loaded CCNPs coated by multilayers (core-shell) structure and surface morphology were confirmed by SEM (Figure. 3.5. D). The core-shell particle exhibited homogeneous spherical shapes with an apparent PELs monolayer on the surface. Morphology of $CaCO_3$ surrounded by layers of PDADMAC, PDCPP appears like a "core-shell" structure.

The obvious difference in the surface morphology of CCNPs and MLNPs confirms the formation of multilayer around CCNPs. The rough surface of the core-shell particle was due to the adsorption of the rigid polyelectrolyte multilayer. The size of the nanoparticles was found to be 250-300 ± 20 nm. The core-shell particle has indeed grown in the size compared to the CCNP score. Conversely, a new polymer layer has covered outside the particle, leading to the further increased size. However, the particles were unstable and agglomerated when fabrication of multiple polymer layers on CCNPs (Figure. 3.5. E). It is important to note that multiple polymer layers (PDADMAC-PDCPP) exhibit more randomly aggregated bulk material with wavy wrinkles.

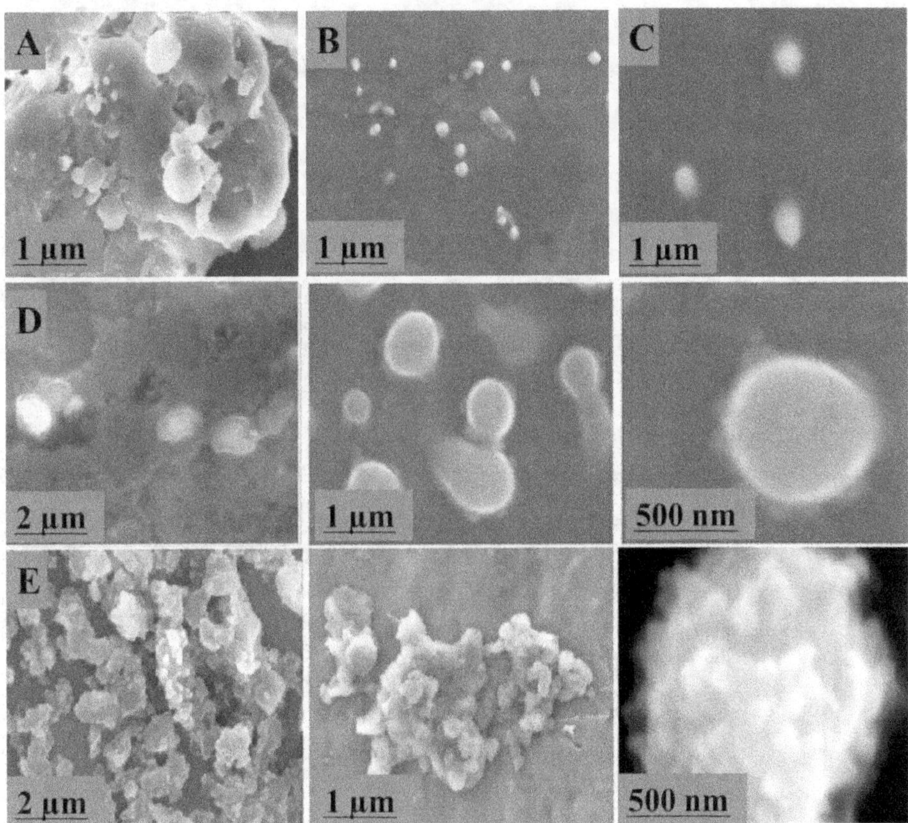

Figure. 3.5. SEM images represent the surface morphology of cisplatin and chrysin-loaded CCNPs (A) CCNPs synthesis by 800 rpm mixing (B) CCNPs synthesis by 1400 rpm mixing (C) CCNPs synthesis by 2000 rpm mixing (D) Demonstrates the high magnification of CCNPs coated with 5 layers of PDADMAC/PDCPP (E) Demonstrates the high magnification of CaCO$_3$ NPs coated with 10 layers of PDADMAC/PDCPP.

Additionally, the TEM image confirms the core-shell structure with non-aggregated, spherical and compact morphology. It is clearly shown that spherical shape surrounded by a layer and dark border inside CCNPs coated with multilayer polymer of 255 in diameter (Figure. 3.6. A, B, C). Although multilayer could not be distinguished clearly because of the longitudinal attachment of polymers.

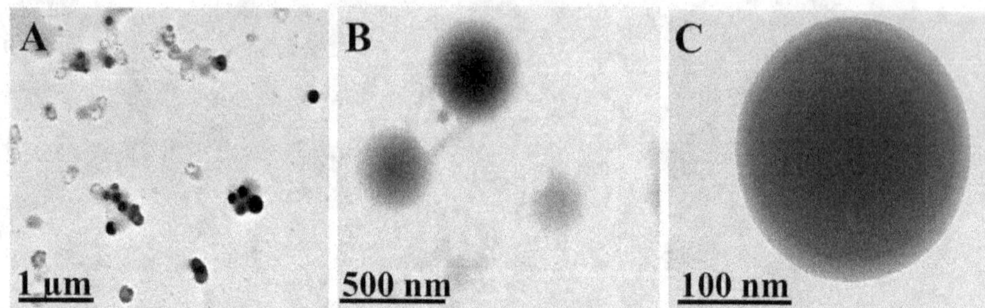

Figure. 3.6. TEM images of CaCO$_3$ NPs coated with 5 layers of PDADMAC/PDCPP (A) 1 μm, (B) 500 nm (C) 100 nm scale bar.

3.6.3. Drug loading and encapsulation efficiency studies

The drug loading and EE of cisplatin/chrysin-loaded CCNPs coated with PDADMAC-PDCPP layers were represented in Figure. 3.7. Reaction time and agitation speed greatly influenced the drug-loading and EE of CCNPs. The cisplatin loading and EE show about 42 and 54 % respectively at 2000 rpm for 12 h (Figure. 3.7. A). Less drug loading and EE were observed at 1400 rpm for 12 h, it might be lower agitation speed. It is found that a considerable quantity of drug loss was observed due to the CaCO$_3$ nucleation process. In Figure. 3.7. B, chrysin loading and EE reached at the maximum of 46 and 60 % respectively for 2000 rpm for 12 h reaction. At greater rpm, the nucleation rate is high which triggering to encapsulation drug shortly. The extended reaction time has also shown decreasing trend of drug-loading and EE in both the drugs.

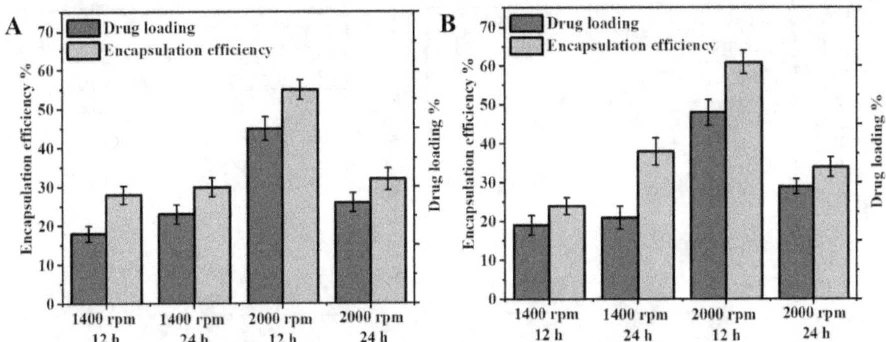

Figure. 3.7. Drug loading and encapsulation efficiency of MLNPs from the encapsulation process and the co-precipitation process, with mixing speed and reaction time (A) Cisplatin (B) Chrysin.

3.6.4. Drug release studies

The drug release profile is based on the solubility, diffusion, and decomposition of the polymer matrix [96]. PDADMAC-PDCPP system around $CaCO_3$ contains a hydrophilic character which enhances water permeation and drug diffusion through the swollen matrix. The release of drugs from the core-shell is primarily controlled by the diffusion-control mechanism. Initially, MLNPs controlled the burst release and followed by exhibits the sustained release of the drug. In addition, it should be noted that the stability of polymer (PDADMAC-PDCPP) during the release study would play a significant role in the controlled release of the drugs. The initial drug release 25 % of the total cisplatin and chrysin was released from the core-shell in acidic pH and 15 % of drug release in neutral pH (Figure. 3.8. A, B). In 80 h, 83 and 84 % of drug release in acidic pH and 62 % of release profile in neutral pH. This result demonstrates that nanomaterial protonated at the acidic pH, which destabilizes the MLNPs and results in the release of the therapeutic load at the desired tumor site.

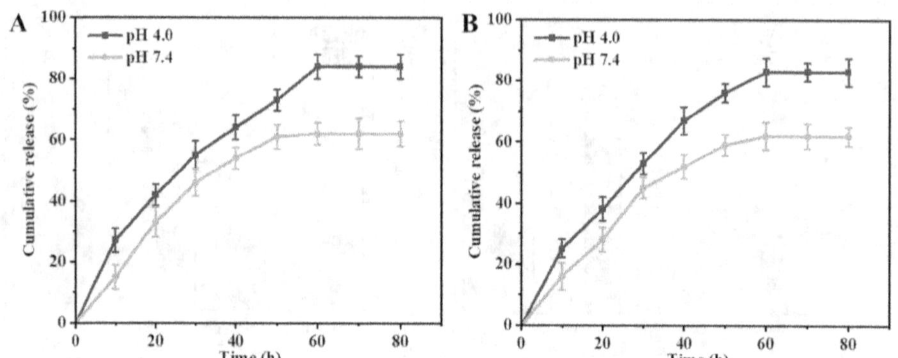

Figure. 3.8. The cumulative release profile of (A) Cisplatin (B) Chrysin from the CaCO$_3$ coated with PDADMAC-PDCPP MLNPs in pH 4.0, 7.4.

3.6.5. *In vitro* cancer studies
(i) MLNPs cytotoxicity studies

The cytotoxic effects of drug-loaded MLNPs on KB cells (oral cancer cells) for 12 and 24 h. The chrysin MLNPs exhibited a mild cytotoxic effect on KB cells showed IC$_{50}$ of about 40 and 20μg.mL^{-1} for 12 (Figure 3.9. A) and 24 h (Figure 3.9. B), whereas the cisplatin and dual drug-loaded MLNPs clearly indicated high cytotoxicity. Cell toxicity relies on time and dose-dependent and generally on high concentration. The IC$_{50}$ of cisplatin MLNPs was 25 and 12μg.mL^{-1} at 12 and 24 h. In Figure 3.9.A, B, both cisplatin and chrysin-loaded MLNPs containing exhibited the high cytotoxicity showing IC$_{50}$ of about 12 and 2 μg.mL^{-1} for 12 and 24 h. This result suggested a decrease in cell viability with a low concentration of cisplatin. A combinational drug helps to eliminate the adverse effect of the cytotoxic drug by reducing the concentration of cisplatin. Interestingly, the synergetic effects of nanocarrier containing cisplatin and chrysin were the most effective formulation against the oral cancer cells compares with individual drug treatment, thereby indicating that the combination therapy potentially possessed a wider application in terms of cancer treatment.

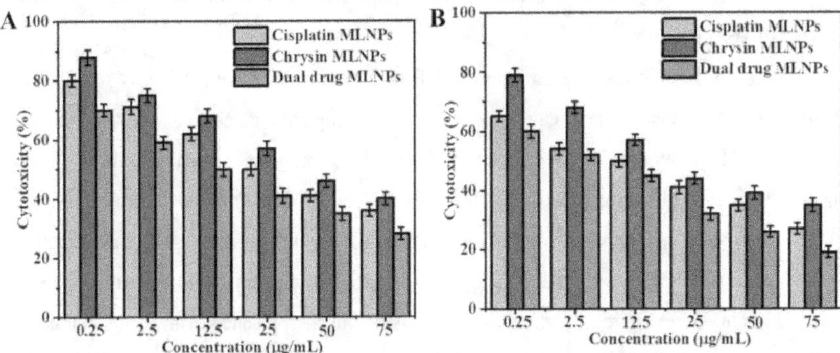

Figure. 3.9. Cytotoxicity analysis after exposure of cisplatin and chrysin-loaded MLNPs to KB cells for (A) 12 h (B) 24 h. All the data are expressed as the mean ±SD of the three experiments with duplicate wells. p<0.05 compared with the control group.

(ii) MLNPs cancer cell penetration studies

The efficiency of cellular uptake and intracellular distribution of MLNPs on KB cells. MLNPs internalized into the cell via passive diffusion and the uptake of MLNPs was increases with the treatment time which compared with control (Figure. 3.10. A). The fluorescence intensity inside the cell was uniform in the cytoplasmic vesicles and even throughout the perinuclear region of the cells after 4 h treatment (Figure. 3.10. B). In Figure. 3.10. C, the image shows the maximum internalization of MLNPs inside the cell exhibits fluorescent sphere shapes and even throughout the perinuclear region of the cells after 8 h treatment [97]. MLNPs are directly internalized via endocytosis without any prior accumulation on the cell membrane. MLNPs penetrate easily into cell membranes due to positive charge and leading to intracellular drug release [98, 99].

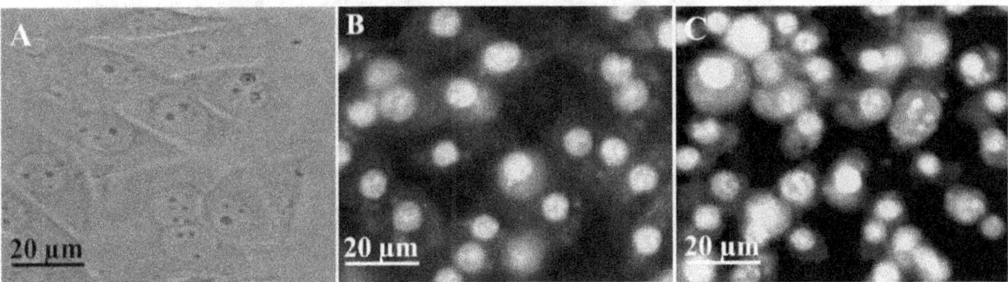

Figure. 3.10. Cellular uptake of FITC loaded MLNPs using KB cells (A) Control cells under Phase contrast microscope (B) After 4 h treatment FITC loaded MLNPs (C) After 8 h treatment FITC loaded MLNPs.

(iii) AO/EB staining

Based on the fluorescence emission in KB cells before stain with AO/ EB staining revealed that cells undergone apoptosis and are categorized into morphological changes and condensation of chromatin in the nuclei. The treatment of KB cells with 25 µg.mL^{-1}of drug-loaded MLNPs for 24 h represent and compared with control (Figure. 3.11. A). The chrysin-loaded MLNPs exhibited mild apoptosis (Figure. 3.11. B). Apoptosis incidence was high in cisplatin-loaded MLNPs, because it was an anticancer drug (Figure. 3.11. C). Hence, it exhibits high apoptosis compared to chrysin. Thus, the MLNPs containing both cisplatin and chrysin exhibited the greatest apoptosis due to the presence of a greater number of necrotic cells (Figure. 3.11. D). Cisplatin or chrysin alone was shown to be the minimum induction of apoptosis compared to dual drug-loaded MLNPs. The synergistic action of both agents was stimulating higher apoptosis in oral cancer cells and more effective in its pharmacological action as compared to free cisplatin/chrysin.

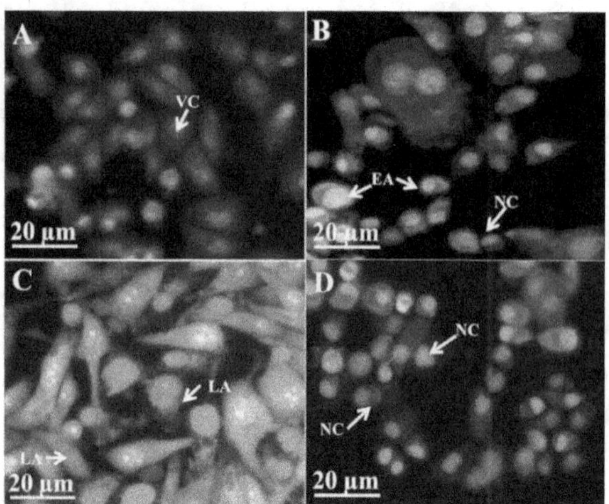

Figure. 3.11. Apoptotic effect of cisplatin and chrysin-loaded MLNPs into KB cells was stained by AO/EB and observed under a fluorescence microscope (A) Control (B) Chrysin MLNPs (C) Cisplatin MLNPs (D) Dual drug loaded MLNPs for 24 h.

(iv) Hoechst Staining

In this study, viable cells exhibit blue nuclei with uniform structure, apoptotic cells consist of irregular nuclei due to the karyopyknosis and nuclear condensation, whereas dead cells nuclei could not be stained [100]. Control cells have typically rounded and light blue nuclei (Figure. 3.12. A). On treatment with MLNPs, most of the cells emitting a high amount of blue fluorescence, and the nuclei of oral cancer cells show hyper condensed nuclei (Figure. 3.12. B, C). Cells treated with chrysin was also observed the formation of apoptotic bodies, confirming the chrysin induces apoptotic cell death. The number of condensed chromatin was increased significantly when cells were treated with dual drug-loaded MLNPs (Figure. 3.12. D). This shows the combined effect of cisplatin and chrysin on apoptosis induction and destruction of the nuclei leading to a severe reduction in the cell population.

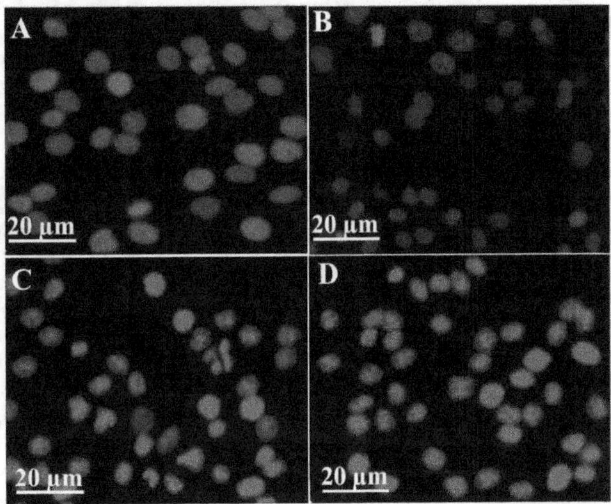

Figure. 3.12. MLNPs effect on KB cells nuclei damage was observed by Fluorescence microscope using Hoechst staining (A) Contol (B) Chrysin MLNPs (C) Cisplatin MLNPs (D) Dual drug loaded MLNPs for 24 h.

(v) Determination of Reactive Oxygen Species

DCFH-DA-stained KB cells showed increased fluorescence intensity when compared to control cells, suggesting that the cisplatin and chrysin loaded in multilayer were triggering to generate ROS (Figure. 3.13. A, B, C). The chrysin-loaded MLNPs show a significant impact on ROS production and because of their antioxidant properties. There was an increase in fluorescence intensity of KB cells treated with dual drug-loaded MLNPs than cisplatin and chrysin-loaded MLNPs (Figure. 3.13. D). This is confirming the combined effect of cisplatin and chrysin in ROS production. In Figure. 3.13. E, the increased levels of ROS play a central role in the regulation of oral cancer cell apoptosis, induces endoplasmic reticulum stress, genomic instability, and cell death [101].

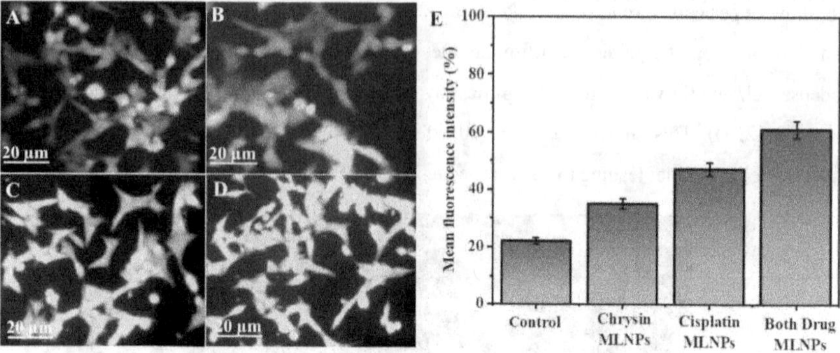

Figure. 3.13. Effect of ROS generation by cisplatin and chrysin-loaded MLNPs into KB cells observed under a fluorescence microscope (A) Control (B) Chrysin MLNPs (C) Cisplatin MLNPs (D) Dual drug loaded MLNPs for 24 h (E) Bar diagaram representation of DCF fluorescence intensity, which is proportional to the amount of ROS produced by the KB cells. The Data represent mean ± SD.

(vi) Measurement of Mitochondrial Membrane Potential

Mitochondrial dysfunction plays a vital role in triggering apoptosis, Rh-123 is a tracer dye and form fluorescent complexes to activate mitochondrial membranes [102], which are employed to determine the loss of mitochondria membrane potential (MMP) in oral cancer cells. The present study showed an increased accumulation of dye in the cells without treatment (Figure. 3.14. A). Figure. 3.14. B, C, represents the chrysin MLNPs with lesser decay of fluorescent complex and cisplatin MLNPs have higher decay of fluorescent complexes. In 25 $\mu g.mL^{-1}$ of dual drug-loaded MLNPs treated cells as revealed by a significant ($P < 0.05$) decrease in MMP (Figure. 3.14. D). Combinational drug treatment effectively induces apoptosis in oral cancer cells by release of cytochrome C, which activates caspase cascade (Figure. 3.14. E). Both drugs loaded MLNPs can enter the cells and distribute drugs rapidly to injure the nucleus and mitochondria more efficiently than a single drug [103].

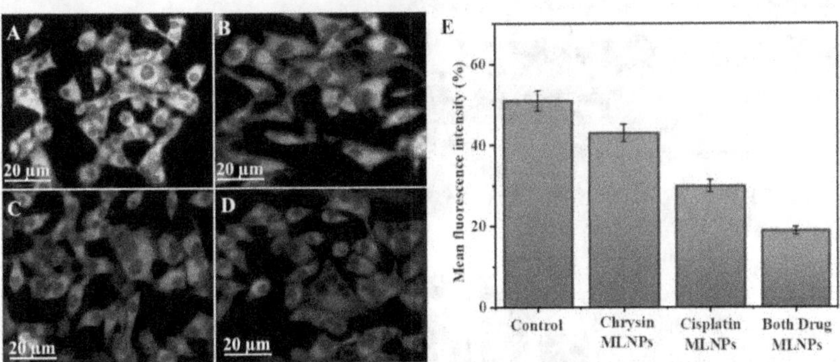

Figure. 3.14. Effect of MMP by cisplatin and chrysin-loaded MLNPs into KB cells observed under the fluorescence microscope (A) control (B) Chrysin MLNPs (C) Cisplatin MLNPs (D) Dual drug loaded MLNPs for 24 h (E) Bar diagaram representation of Rh-123 fluorescence intensity in decreasing manner, which resembles the mitochondrial membrane depolarization on KB cells. The Data represent mean ± SD.

3.6.6. Western blotting

Expression of apoptotic proteins after treatment with oral cancer cells was demonstrated by western blot and it compared with control (Figure. 3.15. A). Single drug-loaded MLNPs of 25 $\mu g.mL^{-1}$ regulates the minimal amount of alteration in protein expression. The upregulation of Bax and caspase-3 was shown by oral cancer cells treated with 25 $\mu g.mL^{-1}$ cisplatin and chrysin-loaded MLNPs, which results in activation of programmed cell death (Figure. 3.15. B, C). Bax protein is a member of the Bcl-2 family, is a pro-apoptotic protein, and plays an important role in apoptosis regulation [104,105]. Dual drug-loaded MLNPs downregulates Bcl-2, leads to activation of caspase and the induction of apoptosis because it releases cytochrome C from the mitochondria (Figure. 3.15. D). The caspases-cascade (cysteine-requiring aspartate proteases) can be triggered by initiators like caspase-8, 9, and effectors caspase-3,7,6 is the most important for induction of apoptosis [106]. This study has pointed out that the activation of caspase-3 has shown a direct effect on the dysfunction of MMP and apoptotic induction. More hydroxyl constituents in the chrysin raise the synergistic effect of a drug and activate the caspase cascade process.

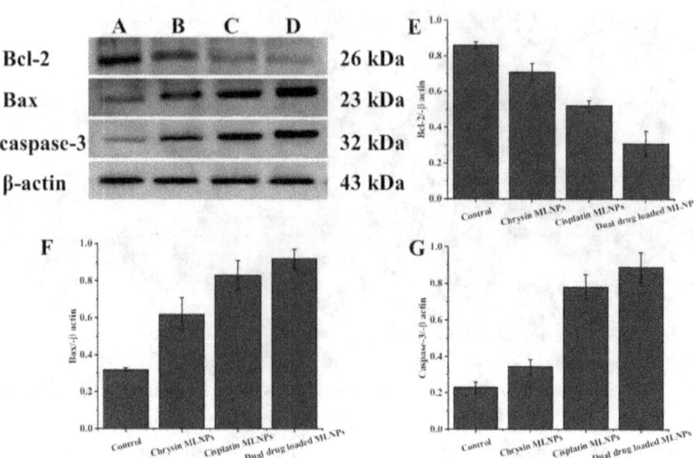

Figure. 3.15. The expression of Bcl-2, Bax, caspase-3 in KB cells after treatment with chrysin MLNPs (A) control (B) Chrysin MLNPs (C) Cisplatin MLNPs (D) Dual drug loaded MLNPs for 24 h was analyzed western blot, respectively (E) The relative protein expression of Bcl-2 which compared with that in the control group (F) The relative protein expression of Bax which compared with that in the control group (G) The relative protein expression of caspase-3 which compared with that in the control group. The whole KB cell extracts were prepared and analyzed by Western blot analysis using antibodies against BCL-2, Bax, and caspase-3. The same blots were stripped and re-probed with a β -actin antibody to show equal protein loading.

3.6.7. *In vivo* antitumor effect of MLNPs
(i) *In vivo* model body weight measurement

The Table. 3.1 data represents the mean body weight of control and experimental animals during the initial and final weeks of the experiments. DMBA alone treated hamsters (group 2) drastically decreased the mean body weight compared to the control group. Oral supplementation of DMBA+cisplatin/chrysin-loaded MLNPs treated animals mean final body weights were significantly increased compared to DMBA alone treated hamsters. MLNPs (group 7) and control (group 1) show no unpredictable changes in body weight. Moreover, dual drug-loaded MLNPs have shown a distinct effect on body weight compared to single drug-loaded MLNPs.

Table. 3.1. Initial and final body weight changes of control and experimental hamsters in each group.

Group	Treatment	Initial body weight (g)	Final body weight (g)
1	Control	116.3 ± 11.23	163.5± 14.76
2	DMBA	118.2 ± 13.32	134.1 ± 31.65
3	DMBA+MLNPs	117.1 ± 9.21	131.3 ± 29.43
4	DMBA+ Chrysin loaded MLNPs	116.4 ± 18.21	149.5±36.75
5	DMBA+ Cisplatin loaded MLNPs	118.3 ± 12.31	152.4±52.69
6	DMBA+ Chrysin &Cisplatin loaded MLNPs	117.2 ± 7.11	157.2 ± 32.87
7	MLNPs	119.5 ± 10.21	161.4 ± 24.65

Data are expressed as mean ± SD for n=6 hamsters in each group.

(ii) *In vivo* tumor regression studies

At the end of the 16th week, the incidence of well-defined exophytic tumor volume (401.65± 13.1 mm^3) and tumor burden (916±75.36 mm^3) was found in DMBA treated animals (group 2) shown in Table. 3.2. No tumor was observed in control and MLNPs alone (group 7) supplemented animals. As expected, cisplatin/chrysin-loaded MLNPs and dual drug-loaded MLNPs administrated with DMBA showed significantly decreased tumor volume and tumor burden in animals. Group 6 (dual drug-loaded MLNPs treated animals) tumor incidence, mean tumor volume, and tumor burden were low compared to groups 3, 4, and 5(cisplatin/chrysin individually loaded MLNPs). The antitumor efficacy of MLNPs was excellent in the accumulation and delivery of dual drugs at the targeted sites.

Nanomaterials are most promising on drug delivery and effectively damages cancer cells compared to the actual clinical protocol for oral cancer [107]. Notably, oral drug delivery methods have been feasible for the treatment of oral cancer due to the ability of the therapeutic to penetrate oral tumor sites and achieve efficiency.

Table. 3.2. Tumor incidence, tumor volume and tumor burden in the control and experimental animal in each groups (n=6).

Group	Treatment	Tumor incidence	Total number of tumors / animals	Tumor volume (mm^3)	Tumor burden	Hyperplasia	Dysplasia
1	Control	0	0	0	0	-	-
2	DMBA	6/6	21/6	401.65± 13.1	916±75.36	+ + +	+ + +
3	DMBA+MLNPs	6/6	19/6	398.22± 15.4	898±34.14	+ + +	+ + +
4	DMBA+ Chrysin loaded MLNPs	5/6 (83.3%)	15/5	309.31±5.4	485±22.11	+ +	+ +
5	DMBA+ Cisplatin loaded MLNPs	4/6 (66.6%)	11/4	289.89± 11.9	373±84.24	+ +	+ +
6	DMBA+ Chrysin &Cisplatin loaded MLNPs	1/6 (16.6%)	1/1	31.67± 41.5	66±12.32	+	+
7	MLNPs	0	0	0	0	-	-

Values are expressed as mean ± standard deviation (S.D.) for six animals in each group.
Mean tumor burden was calculated by multiplying the mean tumor volume ($4/3\pi$) [D1/2] [D2/2] [D3/2] with the mean number of tumors (r=1/2 tumor diameter in millimeter).
- no change, + mild, ++ moderate, +++ severe.

(iii) Biocompatibility of MLNPs using histopathology studies

To obtain an effective nanocarrier for therapeutic applications, it was compulsory to evaluate their potential biocompatibility (Scale bar- 20X). In Figure. 3.16. A, the control group, (without tumor induction) normal tissue keratin layer was in regular thickness and stratified squamous epithelium. This study revealed increased cellular plemorphis, hyperplasia, hypergranulosis, acanthosis was observed in the basal layer, well-differentiated squamous cell carcinoma in DMBA-treated hamsters (Figure. 3.16. B). In addition, there was an irregular extension of the epithelial lesion through the basal membrane into the conjunctive tissue as well as invasive epithelial squamous. In Figure. 3.16. C, D, represents the MLNPs and chrysin loaded MLNPs treatement, without any major change. Drug-loaded MLNPs show less leukocyte count indicates a lack of inflammation and allergic (Figure. 3.16. E). MLNPs exhibit pro-inflammatory potential without any damage and no significant histopathological change. Thus, this result reveals the biocompatibility nature of MLNPs and also the effective treatment of oral cancer (Figure. 3.16. F).

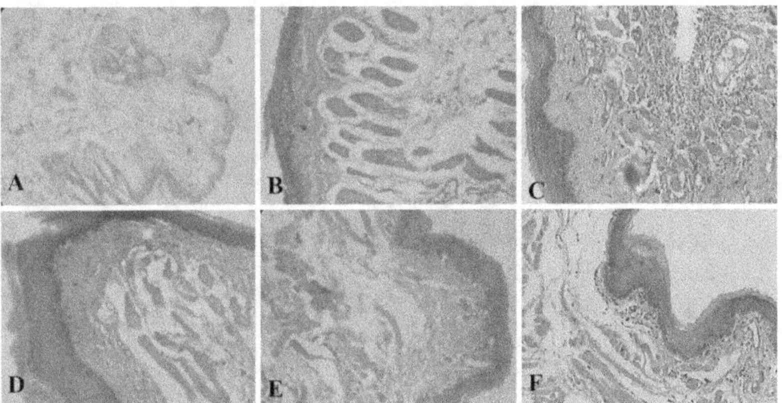

Figure. 3.16. Haematoxylin and Eosin stained tissue sections (A) Control hamsters showing normal epithelium of buccal tissue (B) DMBA alone treated hamsters (C) MLNPs alone + DMBA treated animal tissue (D) Chrysin loaded MLNPs +DMBA treated hamster (E) Cisplatin loaded MLNPs+DMBA (F) Chrysin and cisplatin loaded MLNPs +DMBA treated hamster.

3.7. Conclusion

Multiple anticancer drugs in a single nanocarrier system are a promising therapeutic strategy against oral cancer. MLNPs with cisplatin and chrysin co-encapsulated in CCNPs and covered with alternate layers of PDCPP/ PDADMAC polymer shows good result. The MLNPs are value-added with encapsulation of multiple drugs and cancer-specific drug release. MLNPs shoes good in bio-distribution, sustained release, reduced systemic toxicity, and enhanced therapeutic effectiveness. A strong cellular internalization is exhibited by MLNPs in oral cancer cells. The synergistic effect of the dual drug-induced cell apoptosis, elevates ROS levels, mitochondrial membrane damage, and alters apoptotic protein expression. The combinational drug resulted in a superior therapeutic effect in tumor-induced hamsters by a higher rate of oral tumor regression as compared to cisplatin-loaded MLNPs. The biocompatible MLNPs exhibited superior anticancer activities than that of current practices.

4.1. Introduction

The stimuli-responsive nanomaterials can respond to a specific stimulus, which alters its side-chain/properties of nanomaterial. This property is a promising strategy for delivering drugs against breast cancer and it can easily be designed by controlled hydrophobicity and hydrophilicity side chains of the polymer [108,109]. The fine control of hydrophilic and hydrophobic balance concerning polymer composition can result in response to the temperature, pH, ionic, and redox stimuli [110]. Among various stimuli, temperature and pH stimuli-based properties of hybrid polymers were more related to the breast cancer physiological condition. The pH stands out as a significant factor in the breast cancer environment and it was drastically lower than those in normal tissues. The hybrid polymer could be formed by grafting, block co-polymerization, semi-interpenetrating, and co-polymerization between hydrophilic and hydrophobic molecules.

4.2. Previous work done

Several hybrid polymers with stimuli-responsive behavior were used for the delivery of insoluble drugs. For example, Vinyl alcohol and hydroxy-ethyl methacrylate moieties form hydrophilic-hydrophobic block co-polymer [111], and hydrophobic poly (D, L-lactic acid) diacrylate macromer (PDLLAM) [112]. But there is a lack of biodegradability, biocompatibility, poor diffusion, and swelling properties.

4.3. Novelty

Poly (bis (carboxyphenoxy) phosphazene) has drawn interest for their tunable functionality, like thermoresponsive and pH-responsive drug release [113,114]. By modifying side chains of PCPP with polylactic acid (PLA), cholic acid (CA). It plays a crucial role in controlling the degradation rate of polymers and helps in attaining the gelation properties based on the temperature [115-117]. The aqueous polymer solutions were reversibly transformed into gels with changes in temperature [118]. The fabricated co-polymers contain a hydrophobic/hydrophilic region that can serve as vehicles for the delivery of PTX drugs. The pH-responsive hybrid polymers lead to minimizing its side effect of PTX, increase the EPR effect, and advancing the solubility of the drug.

4.4. Objective

The synthesis of novel cross-linked hydrophobic–hydrophilic hybrid polymer formulated from PLA-CA and PCPP. Physical properties of hybrid polymers, surface morphology, swelling behavior, and hydrolytic degradation of swollen hybrid polymers were investigated. PTX drugs were loaded into the hybrid polymer. The pH and thermoresponsive properties of the hybrid polymer were examined in different pH/temperature conditions. The drug release, cell penetration, cytotoxicity, and western blot were examined in MCF-7 cells.

4.5. Experimental techniques
4.5.1. Preparation of PCPP-PLA-CA

The three different ratios of PCPP: PLA (2:1, 2:2, and 2:3) hybrid polymer was fabricated to get desirable composition. Briefly, 1 % wt of Poly (bis (4-carboxyphenoxy) phosphazene) and PLA (50 mg) were taken in a round bottom flask and dissolved in MES buffer (pH 4.0) under nitrogen atmosphere and stirred for 2 h. Then the flask was placed in an oil bath at 50 °C. After 24 h, the solution was washed 3 times using a centrifuge filter with dehydrated ethanol. The byproduct, PCPP-PLA conjugates were dried in a vacuum oven. Finally, PCPP-PLA (1 %) conjugates were suspended in the MES buffer (0.2M, pH 4.0). Then, CA was added and the mixture was stirred well at 40 °C for about 24 h. The resultant solution was dialyzed against water/methanol mixture for 24 h and then against deionized water for 3 days using a dialysis membrane to remove unbound entities. PCPP-PLA-CA conjugates were dried under a vacuum oven and used.

4.5.2. Synthesis of drug loaded hybrid polymeric nanomaterials

PTX (50 mg) was dissolved in 5 mL chloroform stirred overnight at room temperature. For the water phase, 100 mg of the hybrid polymer was dissolved in 10 mL water for 1 h. The two solutions were mixed in a probe-sonicator for 5 min. Subsequently, the suspension was stirred at 400, 800, 1200, 1600 rpm for 24 h. Nanomaterials were washed twice with water and collected via centrifugation at 8000 rpm for 20 min. Subsequently, they were freeze-dried for 2 days and stored at 4 °C.

4.6. Results and discussion
4.6.1. NMR analysis

The study hypothesized that the pH and thermoresponsive based drug release behavior of the prepared hybrid polymeric nanomaterial depends on a hydrophobic and hydrophilic group. It consists of PCPP (hydrophilic), PLA, and CA (hydrophobic). The PCPP was synthesized by precursor molecule hexachlorocyclo triphosphate by thermal bulk polymerization. Chlorophosphoranimine was reacted with hexachloro cyclotriphosphate to form poly (dichlorophosphazene). Chlorine atoms were replaced by carboxylate ester-containing side groups, by reacting propyl p-hydroxybenzoate to form poly [bis (aryloxy) phosphazene]. NMR spectroscopy of intermediate compound poly [bis (aryloxy) phosphazene] was analyzed. ^{1}H NMR -C-**CH** (8.32, d), -C-**CH** (7.46, d), **CH$_2$**-O (4.2, t), CH$_2$-CO (2.7, t), CH$_3$-CH$_2$ (1.8, m), CH$_2$-CH$_2$ (1.3, t) (Figure. 4.1). ^{13}C NMR CH$_2$- CH$_3$ (22.1), CH$_2$- CH$_2$ (45.3), CH$_2$-O (72.5), CH$_2$-O (78.1), CH=CH (126.7), CH=CH (127.3), CH=CH (129.3), CH=CH (131.6), **C**-CO (139.2), **C**-CO (141.7), **C**-CH$_2$ (147.8), **C**-CH$_2$ (152.5), CO-CH$_2$ (207.1) (Figure. 4.2). The hydrolysis of ester groups in poly [bis (aryloxy) phosphazene] to form Poly (bis (carboxyphenoxy) phosphazene). Polymer products obtained were characterized by means of multinuclear NMR spectroscopy. ^{1}H NMR **C**-COOH (10.12, s), C-**CH** (8.14, d), C-**CH** (7.23, d), O-**CH$_2$** (5.5, t) (Figure. 4.1). ^{13}C NMR **CH$_2$**-O (72), **CH$_2$**-O (77), CH=CH (125.10), CH=CH (128.53), CH=CH (133.73), CH=CH (134.73), **C**-CH$_2$ (141.12), **C**-CH$_2$ (147.23), **C**-COOH (173.3) (Figure. 4.2).

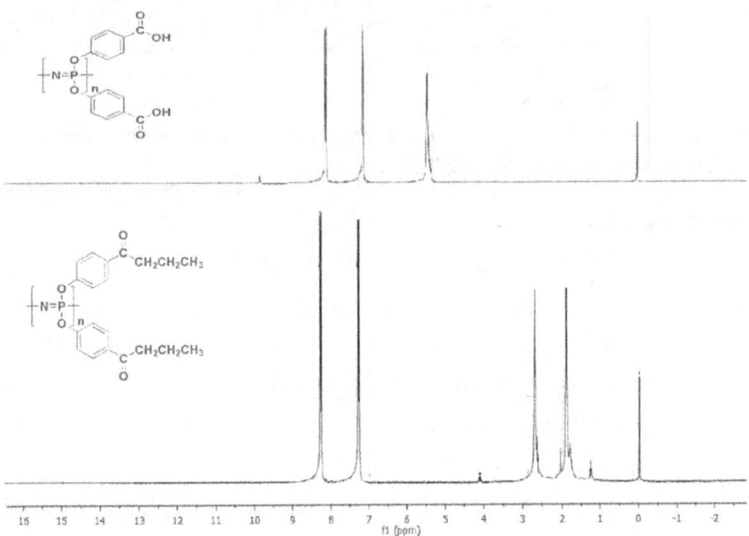

Figure.4.1. 1H NMR spectrum of poly [bis (aryloxy) phosphazene] and Poly (bis (carboxyphenoxy) phosphazene) in DMSO with TMS an internal standard at 300 MHz.

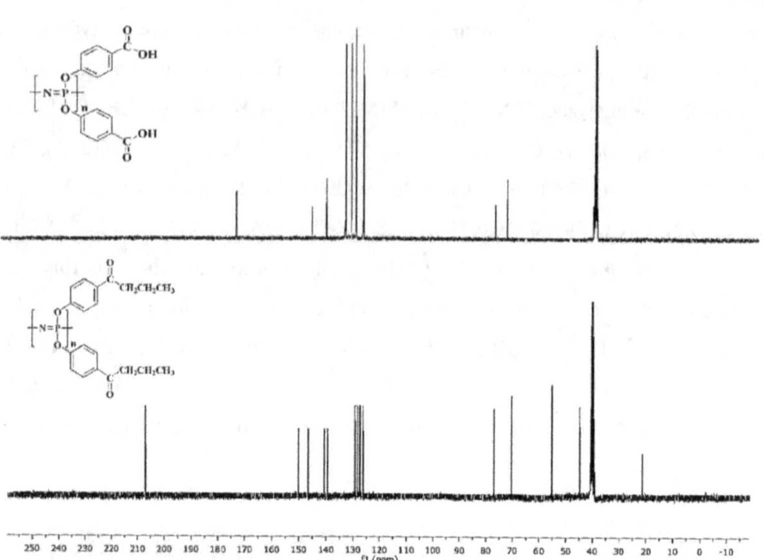

Figure.4.2. ^{13}C **NMR spectrum of poly [bis (aryloxy) phosphazene] and Poly (bis (carboxy phenoxy) phosphazene) in DMSO at 300 MHz.**

4.6.2. FTIR analysis

The synthesis of hybrid polymer (PCPP-PLA-CA) was prepared in three different combinations of PCPP: PLA (2:1, 2:2, and 2:3). The resulting hybrid polymer PCPP: PLA (2:1 and 2:3) was not suitable to use as a carrier. The difference in the ratio combination may affect the mechanical property of the hybrid polymer. Hybrid polymer PCPP-PLA (2:3) has an insoluble and rubber appearance. PCPP-PLA of 2:2 ratios was the optimum concentration to carry the anticancer drug. Furthermore, to observe the functional group characteristics of the co-polymers, the PCPP, PLA, CA, and hybrid polymers subjected to FTIR analysis. The PCPP absorption band corresponds to aromatic C-H, carboxylic acid O-H stretching, aromatic C=C bending, and aromatic C-H vibrations appeared at 3021, 2989, 1774, and 666 cm^{-1}, respectively (Figure.4.3. A). The IR spectra of PLA band at 1081, 3046, 2995 and 751 cm^{-1} correspond to C-O, C-H, -OH stretching, and -C=O bonds vibration (Figure.4.3. B). CA have a very broad band at 3769, 3219 and 2873 cm^{-1} can be attributed to the contribution of H-bond, OH stretch, phenol or tertiary alcohol and 1638, 1463, 1275, 972 and 629 cm^{-1} corresponds to alcohol/ phenol O-H stretch, skeletal C-C vibration, hydroxyl, and aromatic C-H vibration of CA (Figure.4.3. C). The absorption bands of hybrid polymer PCPP-PLA-CA are 3278, 3484, 1519, and 3884 cm^{-1} corresponds to alcohol/ phenol stretch, aromatic C=C bending, aromatic C=C stretch, and -OH stretch. The alcohol /phenol bands represent -OH of CA in the hybrid polymer (Figure.4.3. D). The PCPP polymer consists of a carboxylic acid O-H stretching functional group, which was absent in hybrid polymer. It indicates the formation of PCPP-PLA-CA hybrid polymers.

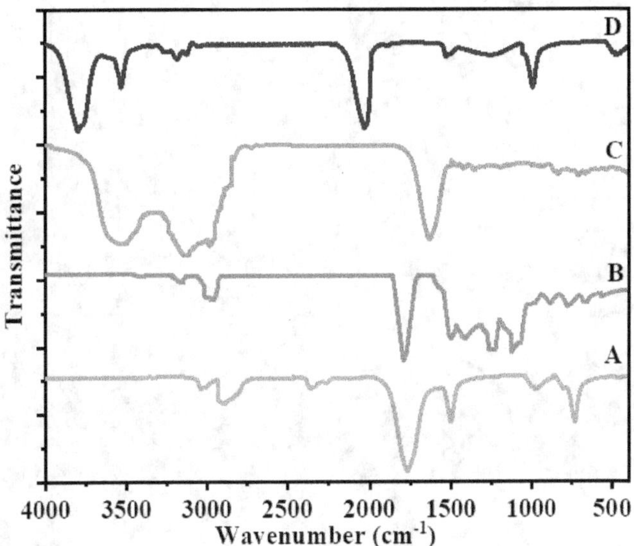

Figure. 4.3.FTIR spectra of (A) PCPP (B) PLA (C) CA (D) PCPP-PLA-CA hybrid polymer.

4.6.3. SEM analysis

Figure.4.4. A, B, C is an unprocessed hybrid polymer that is irregular bulk aggregates with no uniform size. However, when hybrid polymeric nanoparticles exhibited a nanometer size range, predominantly between 300-400 nm (Figure. 4.4. D, E, F). The nanoparticles are fairly uniform in shape, whereas initially polymeric substances are bound together by inter particular bridges and aggregated. This aggregation resulted in poor dispersion of polymers. The synthesized drug-loaded polymeric nanoparticles improved the stability and integrity of polymeric nanoparticles (Figure. 4.4. G, H, I). The surface of the particles was relatively smooth and size between 200-300 nm. Nanomaterials required for slow drug release were prepared by carefully controlling the evaporation of the volatile solvent.

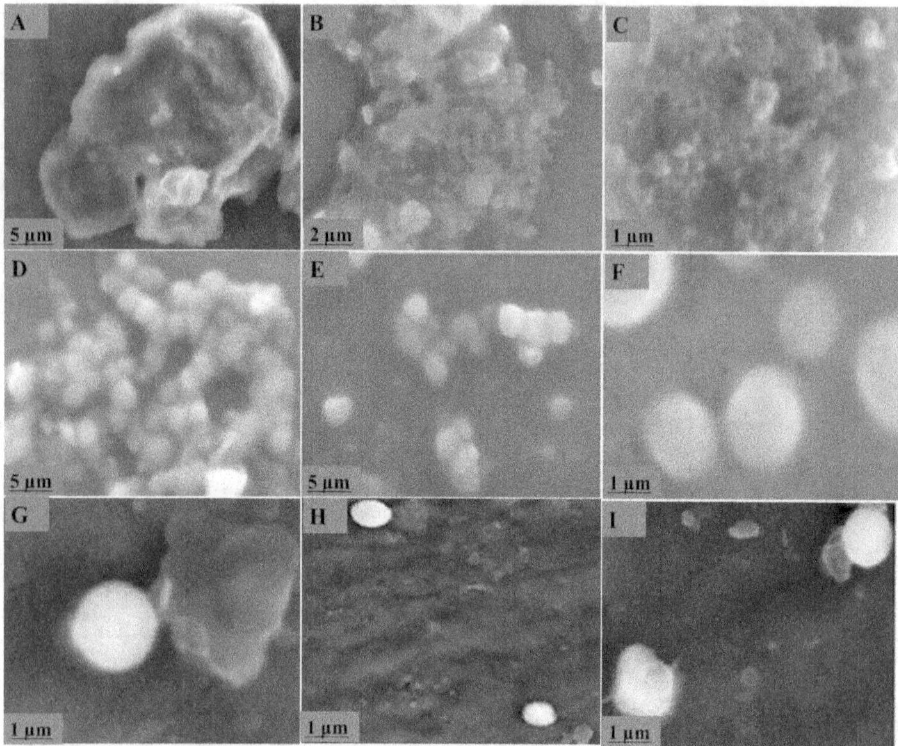

Figure. 4.4. SEM image (A, B, C) Unprocessed hybrid polymer (D, E, F) Hybrid polymeric nanomaterials prepared using oil in water emulsion method (G, H, I) Drug loaded hybrid polymeric nanomaterials.

4.6.4. TEM analysis

The TEM image of hybrid polymeric nanomaterials was depicted in Figure.4.5. A, B, C. respectively presents good agreement with SEM observation. Hybrid polymeric nanomaterials are a spherical structure, smooth surface, and size around 150-200 nm. The large irregular-shaped and merged particles exhibit in the TEM image. These irregular shapes may arise from the merging of hybrid colloidal particles most likely upon solvent evaporation during deposition on grids. The resulting hybrid polymeric nanomaterials are essentially spherical shape, average size, and narrow size distribution.

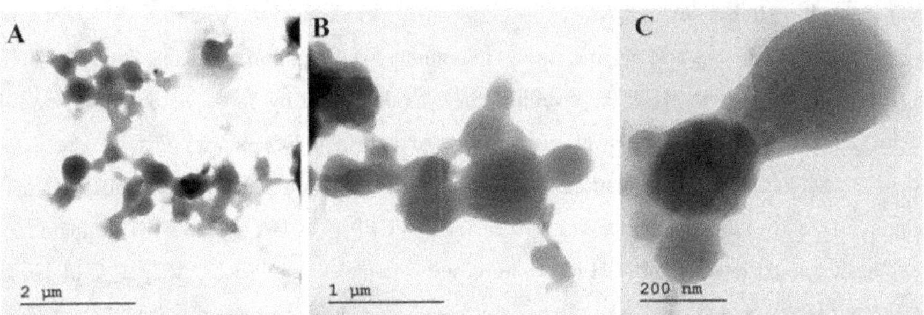

Figure.4.5. TEM image of PTX loaded hybrid polymeric nanomaterial (A) 2 μm, (B) 1 μm, (C) 200 nm.

4.6.5. Thermoresponsive studies

The hybrid polymer solution showed reversible gelation behavior at the temperature between 37 °C and 20 °C. The thermoresponsive behavior of hybrid polymers depends on the temperature-based activation of hydrophilic and hydrophobic constituents. In soft gel, activation of a large number of hydrophilic constituents forms a strong physical junction of hydrogen bonds between the hydrophilic parts in the polymer and water molecules to form an effective liquid hydrogel (Figure. 4.6. B). The hard hydrogel is due to the activation of a large number of hydrophobic constituents and exhibits a high gelation temperature and a high viscosity (Figure. 4.6. A). But gradual temperature increases it becomes opaque and finally starts to shrink by expelling water.

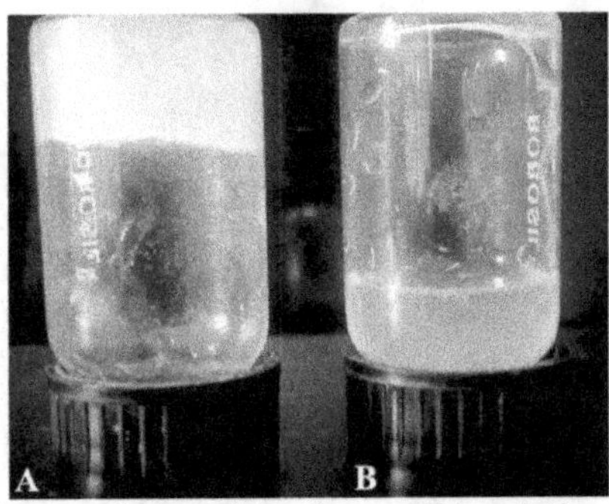

Figure. 4.6. The hybrid polymer solution showed reversible gelation behavior at a temperature between (A) 37 °C (B) 20 °C.

4.6.6. Swelling behavior

The prepared hybrid polymer shows maximum swelling of 80 % for pH 4 and 72 % for neutral pH. Hybrid polymer at pH 9, the swelling rate was decreased by 58 %. Saturation of swelling was achieved after 100 min. The swelling behavior of hybrid polymers was closely related to charge density changes in polymers caused by the functional groups disintegrating at different pH values. Figure.4.7. shows the higher swelling percentage of PCPP-PLA-CA in acidic pH compared to neutral and basic pH. At low pH, the -CH layer swells and becomes more viscous, whereas at higher pH, the -CH layer is elastic and rigid. The larger counter-anions induce the layer to become more viscous and swell to a greater extent than the smaller counter-anions. Swelling of the hybrid polymer involves diffusion of water into the polymeric surface and segmental motion in the interspatial space in the polymer, which ultimately results in an increase of separation distance among polymeric chains [119,120]. Diffusion involves the migration of water into pre-existing or dynamically formed spaces among polymeric chains. The molecules of liquid remain entrapped in these free spaces up to an equilibrium point [121].

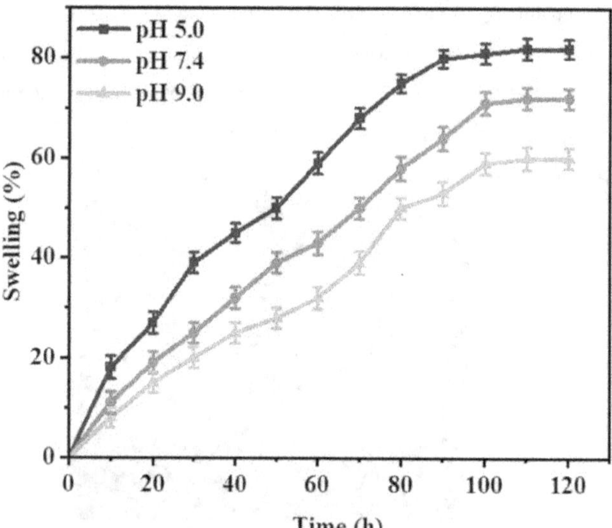

Figure. 4.7. Swelling behavior CA-functionalized PCPP-PLA hybrid polymer in pH 4.0, 7.4, 9.0 at 37 °C.

4.6.7. *In vitro* drug loading and release

The drug loading and EE of PCPP-PLA-CA were represented in Figure. 4.8. The reaction speed greatly influenced the drug-loading and EE of nanomaterials. The higher PTX loading and EE show about 43 and 46 % respectively at 1200 rpm. Less drug loading and EE were observed at 400 rpm and 1600 rpm.

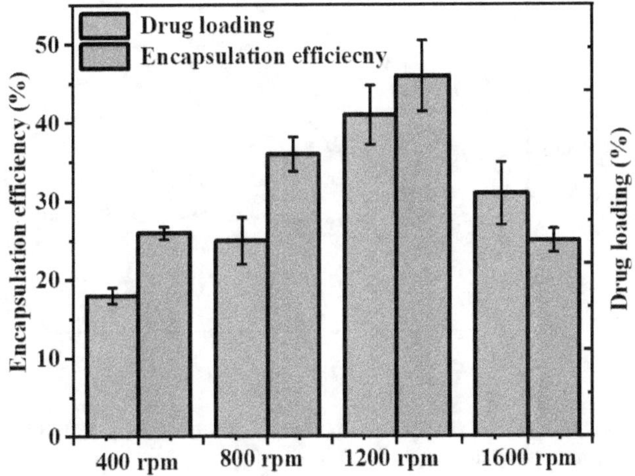

Figure.4.8. PTX loading and encapsulation efficiency of PCPP-PLA-CA.

The PTX-loaded hybrid polymeric nanoparticle was separated from the release medium by a dialysis membrane and studies were carried out at 37 °C at three different pH. The results revealed a significantly faster release of PTX from the hybrid polymer at acidic pH compared to another pH (Figure.4.9). Within 5 days, 54 %, 41% and 37 % PTX were released from polymer at pH 4.0, 7.4 and 9.0, respectively. To achieve 80 % cumulative release, 8 days were required for PCPP-PLA-CA at pH 4.0. The release of PTX from hybrid polymer appeared also highly pH-dependent, in which approximately 90 %, 73 %, and 68 % maximum PTX were released in 12 days at pH 4.0, 7.4, and 11.0 respectively. The drug release was increased as the pH of the media decreased. Hybrid polymer consists of the backbone of the ester bond. The ester bond of hybrid polymer will be easily broken and degraded in an acidic medium (pH 4.0) when compared to the basic medium like pH 7.4 and 9.0. Also, in the acidic condition is accelerated by the protonation and thereby, the drug PTX has been released easily from the hybrid polymer. Drug delivery system maintain higher amount of the drug in the acidic condition which related to the tumor cell environment [122-124]. This study confirms that PCPP-PLA-CA can be potentially used in highly pH-dependent drug delivery.

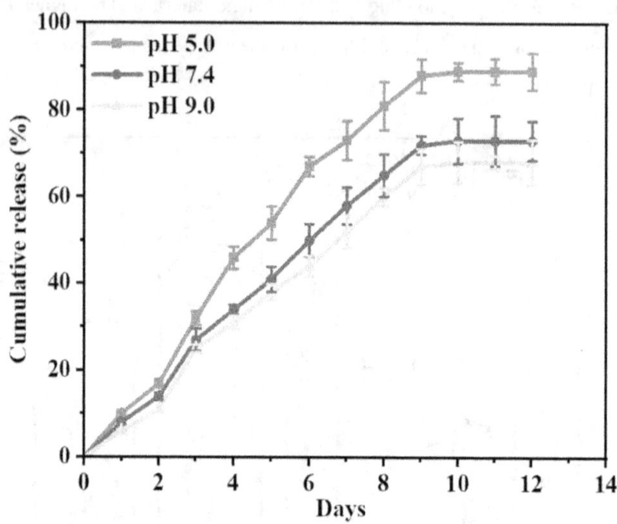

Figure. 4.9. The cumulative release profile of PTX from the of CA-functionalized PCPP-PLA hybrid polymeric nanoparticles in pH 4.0, 7.4, 9.0 at 37 °C.

4.6.8. Cytotoxicity assay
(i) MTT assay

The cytotoxic effects of PTX loaded PCPP-PLA-CA were measured in MCF-7 cells for 24 h. As shown in Figure. 4.10. PCPP-PLA-CA exhibited a mild cytotoxic effect on MCF-7 cells and whereas the PTX indicated more cytotoxicity. Cell viability inhibition was dose-dependent and generally more prominent at higher concentrations. The IC_{50} of PTX loaded PCPP-PLA-CA was 7 µg/mL at 24 h. This result suggests that decrease in cell viability with a low concentration of PTX loaded hybrid polymer. However, the hybrid polymer produced minimal inhibition in MCF-7 cells. PTX inhibits cancer cell growth by induction of apoptosis, cell cycle arrest, and inhibition of angiogenesis, invasion, and metastasis without adverse side effects to normal cells, and the synergistic action of both drugs was required to eliminate the higher breast cancer cells [125,126]. Polymeric drug delivery helps to eliminate the adverse effect of the cytotoxic drug by reducing the concentration of PTX.

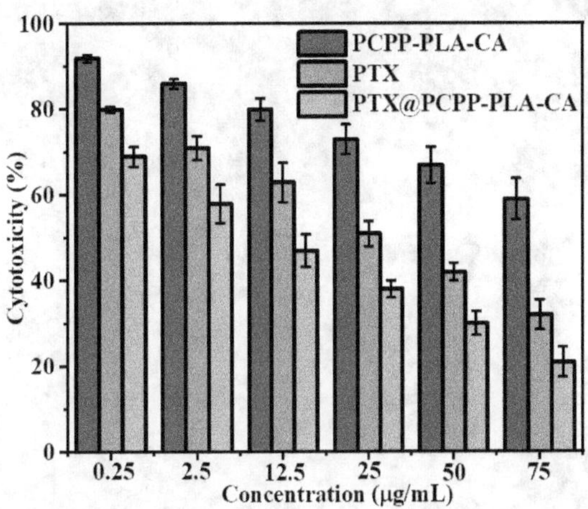

Figure. 4.10. Cytotoxicity analysis after exposure of PTX loaded PCPP-PLA-CA to MCF-7 cells for 24 h. All the data are expressed as the mean ± SD of the three experiments with duplicate wells.

(ii) Apoptosis assay

AO & EB stained treated MCF-7 cells were viewed by fluorescent microscopy revealed apoptosis from the perception of fluorescence emission. Representative Figure. 4.11, shows that the treatment of MCF-7 cells with PTX, PTX loaded PCPP-PLA-CA and drug-free PCPP-PLA-CA nanomaterials for 24 h. Figure.4.11.A shows the more number of live cells. The PCPP-PLA-CA nanomaterials exhibited mild apoptosis (Figure. 4.11. B). Apoptosis incidence was high in PTX loaded PCPP-PLA-CA nanomaterial treatment. So, it exhibits high apoptosis compared to PTX alone (Figure. 4.11. C, D). PTX loaded PCPP-PLA-CA exhibited the greatest apoptosis which was confirmed by a greater number of necrotic cells. PTX alone was shown to be the minimum induction of apoptosis compared to the drug-loaded hybrid polymer. The synergistic action of nanomaterial was stimulating higher apoptosis in MCF-7 cells and more effective in its pharmacological action as compared to free PTX.

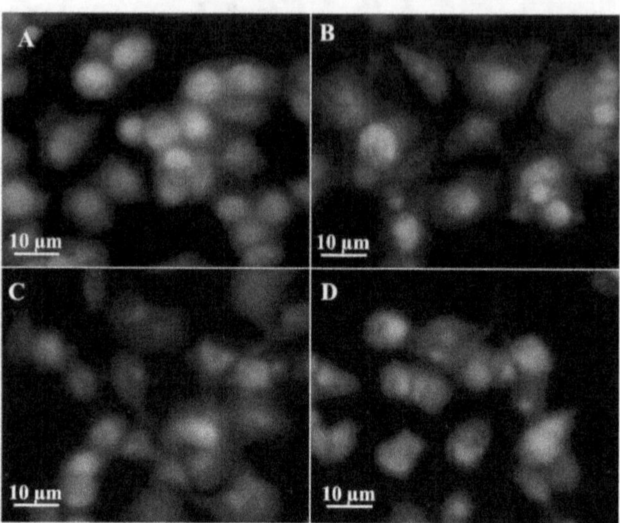

Figure.4.11. Apoptotic effect of PCPP-PLA-CA into MCF-7 cells observed under a fluorescence microscope (A) Control (B) PCPP-PLA-CA (shows early apoptotic stage) (C) PTX (few cells at the late apoptotic stage) (D) PTX loaded PCPP-PLA-CA treated cells and stained with AO/EB staining.

(iii) Measurement of Mitochondrial Membrane Potential

MMP changes were analyzed after treatment with PTX, PTX loaded PCPP-PLA-CA, and drug-free PCPP-PLA-CA nanomaterials on MCF-7 cells for 24 h by Rh-123 fluorescence dye. Mitochondrial dysfunction plays a vital role in triggering apoptosis. Rh-123 is a tracer dye and it can form fluorescent complexes to activate mitochondrial membranes, which are employed to determine the loss of MMP in MCF-7 cells. The rate of fluorescent complex degradation of PTX is higher compared to the control and PCPP-PLA-CA (Figure. 4.12. A, B, C). Treatment of MCF-7 cells with PTX loaded PCPP-PLA-CA causes a decrease in $\Delta\psi$m owing to mitochondrial membrane depolarization, which confirmed the activation of mitochondria-mediated apoptosis (Figure. 4.12. D). Fluorescence microscopic images show increased accumulation of dye in the control compared to drug-loaded nanomaterials. A number of decay fluorescent complexes was observed in PTX loaded PCPP-PLA-CA treated cells as revealed by a significant ($P < 0.05$) decrease in MMP. PCPP-PLA-CA drug treatment effectively induces apoptosis in cancer cells by release of cytochrome C. MMP play an important role in the induction of apoptotic pathway by cytochrome C for activation of the caspase cascade [127,128].

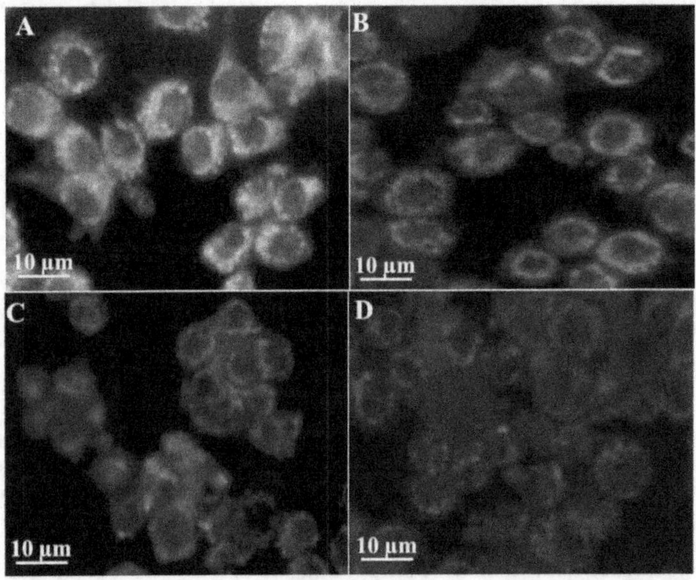

Figure.4.12. Effect of mitochondrial membrane potential (A) Control (B) PCPP-PLA-CA (C) PTX (D) PTX loaded PCPP-PLA-CA treated cells and stained with Rh-123.

4.6.9. Western blotting

Expression patterns of Bax, Bcl-2, caspase-3, and ß-actin were observed after treatment with PCPP-PLA-CA, PTX, PTX loaded PCPP-PLA-CA nanomaterial by western blot analysis (Figure. 4.13. A, B, C). Normally in cancer cells, the caspase pathway is inactive and an anti-apoptotic protein Bcl-2 is over-expressed and it characterizes the production of anti-apoptosis factors in the cells. Bax protein is a member of the Bcl-2 family, is a pro-apoptotic protein, and plays an important role in apoptosis regulation. As shown in Figure. 4.13. D, the upregulation of Bax and caspase-3 was shown by MCF-7 cells after treatment with PTX loaded PCPP-PLA-CA, whereas Bcl-2 was downregulated, which results in activation of programmed cell death. All the expressions were normalized to the levels of the ß-actin expression. The result indicates the downregulation of Bcl-2 leads to activation of caspase and the induction of apoptosis because it releases cytochrome C from the mitochondria. Bcl-2 heterodimerization with Bax, when the Bcl-2 expression level was downregulated, there was upregulation of Bax expression, and homodimers of Bax will always be formed and apoptosis will be stimulated [129,130].

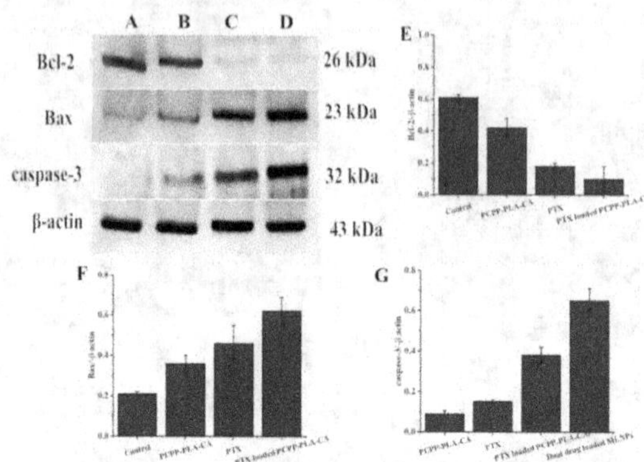

Figure. 4.13. The expression of Bcl-2, Bax, caspase-3 in MCF-7 cells (A) control (B) PCPP-PLA-CA (C) PTX (D) PTX loaded PCPP-PLA-CA nanomaterial for 24 h was analyzed western blot, respectively. (E) The relative protein expression of Bcl-2 which compared with that in the control group (F) The relative protein expression of Bax which compared with that in the control group (G) The relative protein expression of caspase-3 which compared with that in the control group. The same blots were stripped and re-probed with a β -actin antibody to show equal protein loading.

4.7. Conclusion

This study shows that hybrid polymer PCPP-PLA-CA is the suitable new class of polymer for PTX delivery against breast cancer. PCPP was synthesized from the thermal bulk polymerization of hexachlorocyclotriphosphazene and it was predicted by NMR. The biodegradable hybrid polymer was formulated via PCPP (hydrophilic) and PLA, CA (hydrophobic). The polymer synthesized in a controlled manner exhibits desired swelling and hydrolytic properties. Hydrophobic drug PTX was successfully encapsulated with the hybrid polymer to form the spherical nanomaterial. It exhibits reversible gelation behavior at the temperature of 37 °C and 20 °C. This indicates the nanomaterial is able to form a gel at 37 °C and release the drug in a sustained manner. The PTX containing hybrid polymer releases the PTX in an acidic environment of the breast tumor region. PCPP-PLA-CA hybrid polymer makes a promising anticancer effect against MCF-7 cells. It induces apoptosis, forms necrotic cells, and affects the MMP. So, both pH/temperature stimuli-responsive hybrid polymer performs effective therapeutic nanocarrier against breast cancer.

5.1. Introduction

The major breast cancer therapeutic options are chemotherapy, surgery, hormonal therapy, and radiotherapy. But, it show poor efficiency, harmful adverse effects because of their lack of site-specific targeting [131,132]. Polymeric nanomaterials play a great role in anticancer drug development, by providing effective accumulation of nanomaterial in the tumor site. It increases the drug efficiency and anticancer effect. The polymeric nanomaterial charge reversal property and active targeting of breast cancer cells will overcome the above drawbacks of conventional treatment. Charge reversal property of nanomaterial is an alteration of nanomaterial surface charge based on the environmental pH [133,134]. In a neutral environment, the nanomaterial remains a negative charge which could avoid nonspecific binding in normal cells. It became positively charged after reaching an acidic tumor extracellular environment which enhances their cellular uptake [135]. The presence of ligands on the surface of polymeric nanomaterial helps easy internalization and higher accumulation of nanomaterials into cancer cells by receptor-mediated endocytosis [136,137].

5.2. Previous work done

Ligands such as folic acid, hyaluronic acid, peptides, aptamers, and mannose effectively increase the uptake efficiency and therapeutic effect of the drug [138-140]. CA an amphiphilic steroid molecule has the ability to high internalization into breast cancer cells due to strong affinity on farnesoid X receptor (FXR) [141,142]. The CA binds to this receptor on the surface of breast cancer and down-regulates the protein essential for breast cancer cell proliferation [143].

5.3. Novelty

Poly (bis (carboxyphenoxy) phosphazene) (PCPP) controls leakage of a drug in blood vessels, minimizes the release of drugs in normal tissue [144,145]. The targeted drug release at the cancerous site was achieved through CA, it facilitates the enhanced permeability and retention effect in solid breast tumors [146,147]. The charge reversal behavior of polymeric nanomaterial improves breast cancer cell penetration.

5.4. Objective

The designed nanomaterial consists of a CA ligand with PCPP and poly (diallyldimethylammonium chloride) (PDADMAC) polymer. The CA functionalized with PCPP and PDADMAC from polymeric nanomaterials, loaded with PTX. The polymeric nanomaterials, morphology, and particle sizes were characterized by DLS, zeta potential, and microscopic analysis.

The stability, kinetic drug release behaviors, and antitumor efficacy of the hybrid polymeric nanomaterials were investigated. CA binds to hybrid polymers would provide cell-specific targeting pave for precise drug delivery and enhanced therapeutic effect. The charge reversal nature of polymeric nanomaterials enhances the bioavailability of the drugs and helps to reach the cancer cell nucleus. The effective cell penetration and enhanced drug accumulation inside the cancer cell were identified by comet assay and γH2AX assay.

5.5. Experimental techniques
5.5.1. Synthesis of CA grafted PCPP conjugates

PCPP of 1 % wt and CA (50 mg) were resuspended in pH 5.0 MES buffer. EDC (0.10 mM) and NHS (0.30 mM) linkers were added and stirred for 2 h to activate the carboxylic group. Then, the resulting solution was added with 2 mL of ethylenediamine. After 24 h, 2 mL of 50 mM hydroxylamine (pH 7.0) in PBS was added to the solution for another hour to backfill any unbound NHS esters. Last, the solution was washed 3 times using a centrifugation 8000 rpm with dehydrated ethanol to remove unbound entities.

5.5.2. Preparation of CA functionalized PDADMAC

CA-functionalization to PDADMAC by ionic interaction was carried out as per the previous report [148].PDADMAC of 4 % solution was added to the sodium cholate solution. The reaction products were purified by precipitation in ethanol. Further, the precipitates were dialyzed against water for 3 days using a dialysis membrane to remove unbound entities. The resulting polymer was washed and dried using a lyophilizer.

5.5.3. Preparation of CA-PCPP-PDADMAC-CA nanomaterials

The nanomaterials were prepared by the typical procedure [149], CA- PDADMAC of 100 mg was dissolved in 5 mL chloroform for 12 h. CA-PCPP of 100 mg was dissolved in 10 mL water for 2 h. The two solutions were mixed using a probe-sonicator for 10 min. Subsequently, the suspension was kept stirring and allowing complete evaporation of chloroform for 24 h. The final product was washed twice with water and collected via centrifugation at 8000 rpm for 20 min. Subsequently, they were freeze-dried for 2 days and stored at 4 °C.

5.5.4. Comet assay

Briefly, 30,000 cells per well were seeded in a 12 well plate and allowed to gain 70-80 % confluency. MCF-7 cells were treated with PTX loaded polymeric nanomaterials of IC_{50} concentration. After the incubation cells were harvested, mixed with 75 µL of molten agarose, and was spread onto the grooves of comet slides. The slides were kept on ice in dark for 2 h and then immersed in pre-chilled lysis solution and incubated on ice in dark for another 1 h. The slides were then incubated in the alkaline electrophoresis buffer for 45 min in dark. The samples were electrophoresed at 20 V for 15 min. The samples were then washed with chilled distilled water twice and were fixed with pre-chilled 70 % ethanol. After fixation, the samples were stained with 75 µL of SYBR green and incubated in dark for 30 min followed by observing the slides under a fluorescence microscope. "TriTech CometScoreTM Freeware v1.5" software was used to measure the comet tail length.

5.5.5. γH2AX Assay

Cells were exposed to polymeric nanomaterials of IC_{50} concentration at different incubation times. After treatment, cells were washed with PBS and fixed with methanol: acetone (1:1) for 20 min at -20 °C. After fixation, cells were blocked with blocking solution and treated with 10 Ab (anti-γH2AX Ab). After removing unbound antibodies, secondary anti-mouse IgG (anti γH2AX) conjugated to TRITC) was added and incubated for 1 h at room temperature. After incubation, cells were washed thrice with PBS, counterstained with DAPI and the images were captured using 20X magnification of Evos Fluorescent Microscope.

5.6. Results and discussion

5.6.1. Preparation of CA-PCPP-PDADMAC-CA nanomaterials

The CA functionalized polymeric nanomaterial was prepared for breast cancer targeting and stimuli release. The carboxyl group of CA was activated using NHS/EDC, followed by the amide linkage of ethylenediamine to yield CA-PCPP. And also, sodium cholate was formed through ionic interaction with PDADMAC. The CA conjugations of polymers were shown in the ^{1}H, ^{13}C NMR spectra (Figure.5.1, 5.2). The amino carboxyl ratio of CA-PCPP was confirmed by the signals NMR chemical shifts.

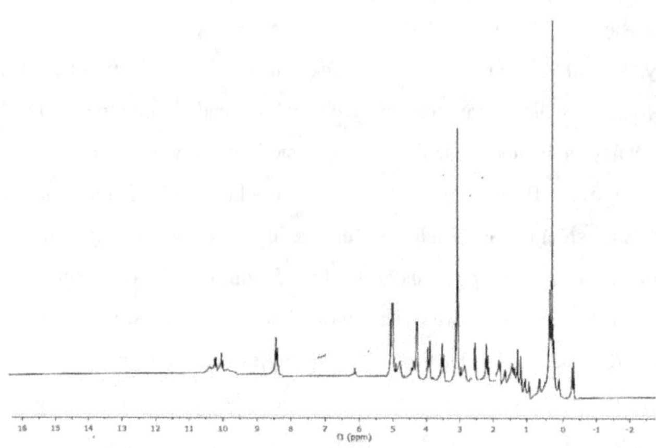

Figure.5.1. ^{1}H NMR spectrum of CA-PCPP in DMSO at 300 MHz.

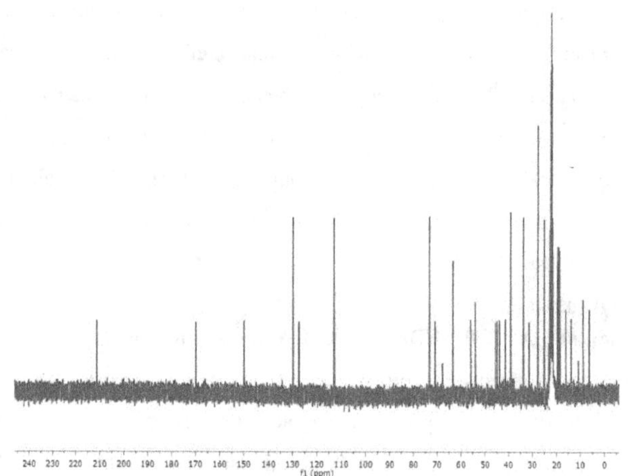

Figure.5.2. ^{13}C NMR spectrum of CA-PCPP in DMSO at 300 MHz.

5.6.2. Characterization of CA-PCPP-PDADMAC-CA
(i) FTIR analysis

CA functionalized onto PCPP, PDADMAC, and drugs incorporated into polymeric nanomaterials have been examined by FTIR analysis. The spectrum as shown in Figure.5.3. A exhibits several adsorption bands at 3008, 1697, 773, 613 cm^{-1} which was assigned to –OH, C=O stretch, C=C stretch denotes the CA grafted on PCPP polymer. On the other hand, CA functionalized PDADMAC shows characteristics peaks at 3003, 2997, 2573, 1082, 797, 676 cm^{-1} which comes from -OH stretch, carboxyl group, C-H, C=C bending, aromatic -CH stretching of cholic acid (Figure.5.3. B). CA hydroxyl and carboxyl group vibrational stretches indicate the proper conjugation of CA to the PDADMAC polymer. The spectrum of CA-PCPP-PDADMAC-CA confirmed the presence of CA in the polymeric nanomaterial. Distinctive new adsorption peaks at 3512, 3001, 1759 cm^{-1} are denoted N-H stretch, -OH, C=O group CA-PCPP-PDADMAC-CA polymer (Figure.5.3. C). Upon comparing the CA-PCPP and CA-PDADMAC transmittance, it was implicated that the FTIR spectrum of CA-PCPP-PDADMAC-CA shows substantial N-H stretch of ethylenediamine, the hydroxyl group of CA, and carboxyl functional groups. In drug encapsulated polymeric nanomaterial were observed without any obvious change (Figure.5.3. D). This observation was due to the successful incorporation of drugs without chemical modification after the nanomaterial formulation.

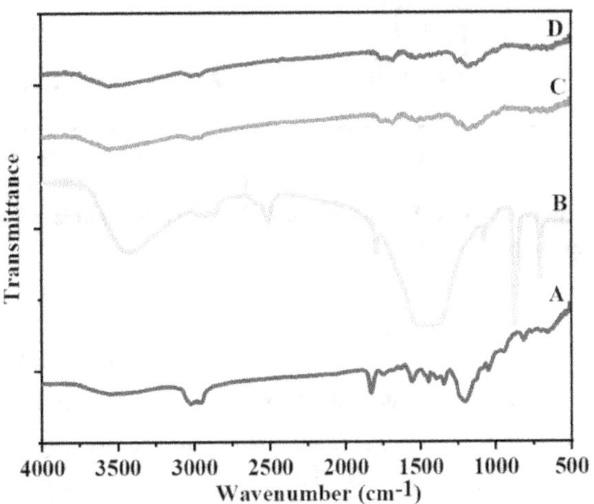

Figure. 5.3. FTIR spectroscopy of (A) CA conjugated PCPP (B) Sodium cholate conjugated PDADMAC (C) Polymeric nanomaterials (D) PTX loaded polymeric nanomaterials.

(ii) XRD pattern

In Figure.5.4, XRD patterns of CA-PCPP have their characteristic peaks at 2θ =14 and 20° due to their semi-crystalline nature. For CA-PDADMAC, the XRD curve forms a broad peak indicateing the existence of amorphous nature. Whereas in CA-PCPP-PDADMAC-CA intensities peaks resemble the crystalline peak of CA-PCPP which indicates that the grafting has not interfered with crystalline nature. And also, the broad peak of CA-PDADMAC was shifted due to the transformation of amorphous to semi-crystalline nature.

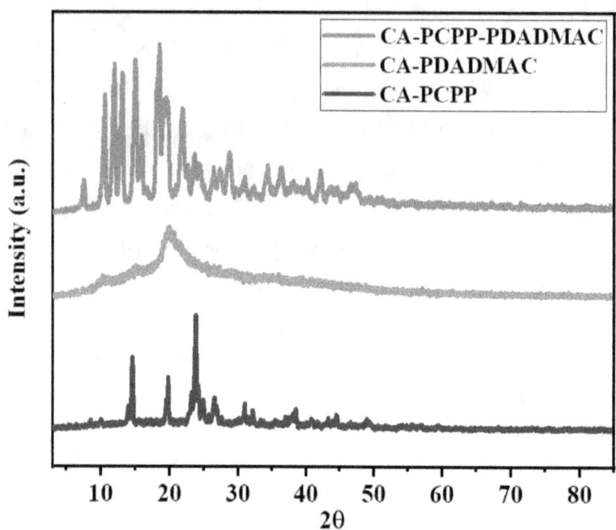

Figure. 5.4. XRD pattern of CA conjugated PCPP, sodium cholate conjugated PDADMAC, polymeric nanomaterials.

(iii) TGA studies

The thermal stability of CA conjugated polymeric nanomaterial was studied via TGA. All grafted copolymers showed similar degradation patterns as demonstrated in Figure.5.5. CA-PCPP degraded rapidly once the temperature reached 130 °C and the majority of the weight loss of CA-PCPP was above 250 °C. The starting degradation temperature of CA-PDADMAC was close to 230 °C. At this temperature, 70 % of CA-PCPP polymer was degraded. The results indicated that CA-PDADMAC was more stable to heat than CA-PCPP, but CA-PCPP-PDADMAC-CA was less stable than CA-PDADMAC. Thermo-stability of CA-PCPP-PDADMAC-CA was enhanced by the surface decoration of CA-PDADMAC. An increase in temperature stability of CA-PCPP-PDADMAC-CA was due to the residual charring product of degraded polymers on the surface.

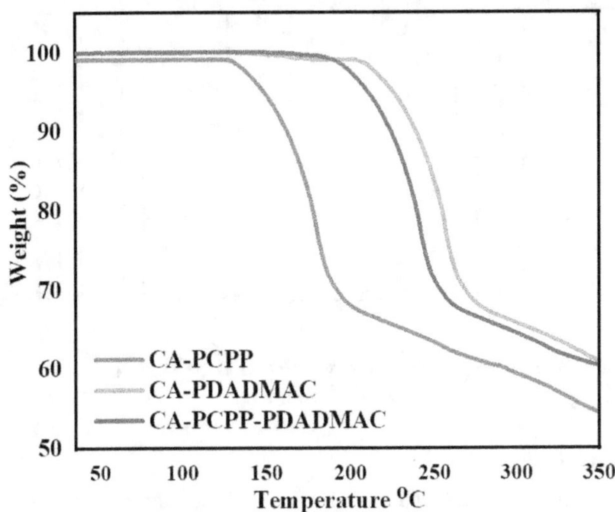

Figure. 5.5. Thermogravimetric analysis curve of CA conjugated PCPP, sodium cholate conjugated PDADMAC, polymeric nanomaterials.

(iv) DLS analysis

The size distributions of polymeric nanomaterials were analyzed using DLS. The empty nanomaterials were of an average size of 218 nm and a polydispersity index of 0.532 (Figure. 5.6. A). The sizes were very similar to those of PTX loaded nanomaterial of 221 nm and polydispersity index of 0.612 respectively (Figure. 5.6. B). Molecular assemblies of CA-PCPP and CA-PDADMAC control the diameter, morphology, and polydispersity of polymeric nanomaterials. The finely regulated polymers concentration makes spherical morphology with 100-200 nm diameters of nanomaterials. Nanocarriers of 100-200 nm size range show higher intracellular uptake with higher enhancing permeation and retention effect [150].

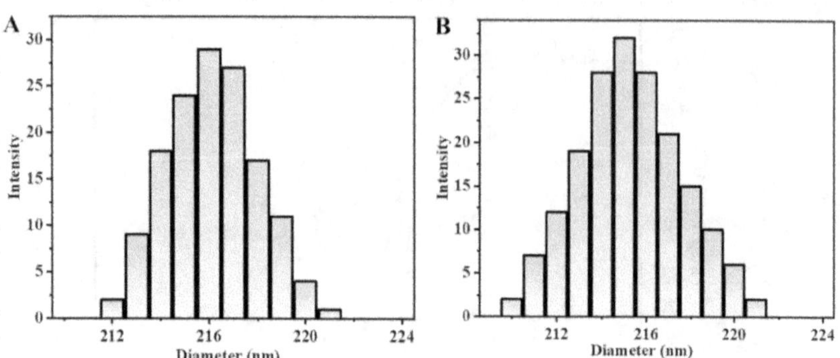

Figure. 5.6. (A) DLS histogram of CA-PCPP-PDADMAC-CA (B) PTX loaded CA-PCPP-PDADMAC.

(v) Zeta potential studies

The Zeta potential of CA-PCPP-PDADMAC-CA nanomaterials undergoes a charge reversal from negative to positive (Figure.5.7). The zeta potential of empty nanomaterial and drug-loaded shows -6.21 and -7.34 mV at pH 7.0 respectively. Interestingly, nanomaterial charge transitions from negative to positive (4.16 mV) at pH 5.0. The positively charged transition of a polymeric nanomaterial at an acidic microenvironment increases higher cellular uptake in cancer cell acidic milieu. The amino/carboxyl ratio of CA ethylene diamine-PCPP increases the charge conversion property of CA-PCPP-PDADMAC-CA. Additionally, negative charge nanomaterial reduces the non-specific uptake of normal cells with neutral charge due to the electrostatic repulsion [151].

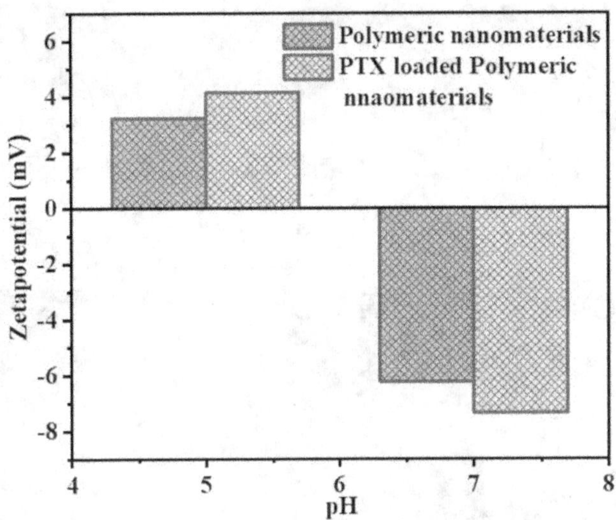

Figure. 5.7. Zeta potential of CA-PCPP-PDADMAC and PTX loaded CA-PCPP-PDADMAC under different pH.

(vi) SEM analysis

Morphology of material is important for strongly increasing the availability of nanomaterial at specific sites. Grafted polymers and polymeric nanomaterials structures were investigated by SEM technique. Polymeric nanomaterials have a modification in morphology compared to the functionalized polymer. CA-PCPP-PDADMAC-CA polymeric nanomaterials form the spherical shape with a 100-200 nm size range (Figure.5.8. A, B). Nanomaterials formation and chemical composition are the main reasons for spherical shape. Cholate functionalized polymers are aggregated, bulk material and nanomaterials fabrication with surfactant was produced nanoscale size by mixing of polar/nonpolar molecules. PTX encapsulated CA-PCPP-PDADMAC-CA polymeric nanomaterials also form spherical morphology and the main driving system for shape (Figure.5.8. C, D). Monodispersed nanomaterials structure forms with a diameter in a range of 176 nm with high stability. Polymeric nanomaterials have a high hydrophilic segment which is the reason for stability. The size of the nanomaterials is the key parameter that affects cellular internalization, tumor targeting, and also nanomaterial's interaction with cells. Nanomaterials of 150-200 nm in size were internalized easily in cells and they accumulated in tumors more efficiently [152].

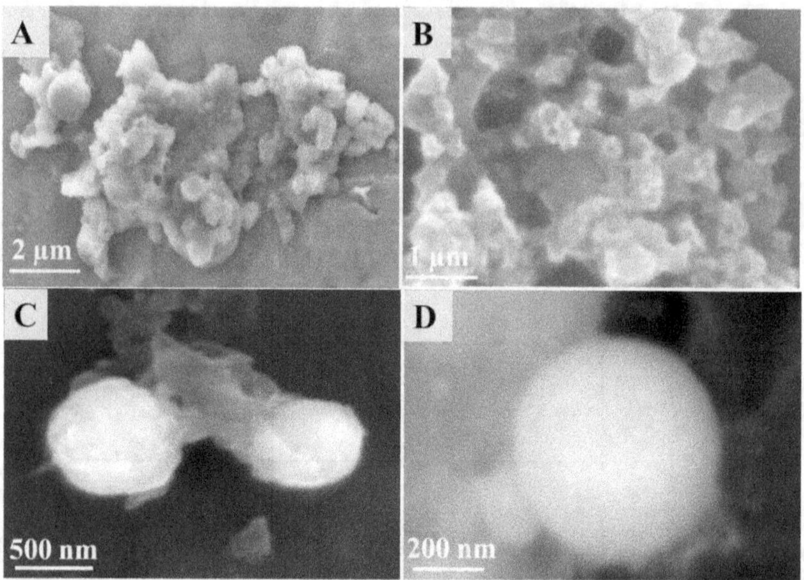

Figure. 5.8. SEM image of (A) CA-PCPP-PDADMAC-CA polymer (2 µm) (B) 1 µm (C) PTX loaded CA-PCPP-PDADMAC-CA (500 nm) (D) 200 nm.

(vii) AFM analysis

An AFM image of CA-PCPP-PDADMAC-CA polymeric nanomaterials and PTX loaded polymeric nanomaterials. Spherical morphology was observed with a rough surface for CA-PCPP-PDADMAC-CA polymeric nanomaterials (Figure.5.9. A). The rough surface of nanomaterials was essential for the binding of nanomaterials on the surface of the cell membrane. CA-PCPP-PDADMAC-CA has exhibited improved entrapment efficiency of PTX drugs. Encapsulation of PTX was not affect the surface morphology (Figure.5.9. B). AFM 3D image of both nanomaterials shows the variation of height indicates the rough surface. AFM tip-broadening effect results in an overestimation of particle size. Nanomaterials of block-co-polymers with hydrophilic and hydrophobic units of CA-PCPP, PDADMAC-CA that self-assembled to spherical structures composed of a hydrophobic core stabilized by a hydrophilic segment. The stable structures confirm the hydrophilic interaction between polymeric nanomaterials.

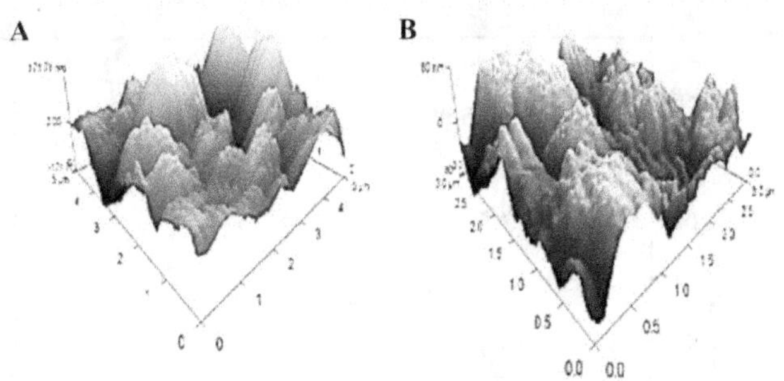

Figure. 5.9. AFM 3D images of (A) CA-PCPP-PDADMAC-CA polymeric nanomaterials (B) PTX drug loaded CA-PCPP-PDADMAC-CA polymeric nanomaterials.

5.6.3. Swelling studies

The swelling profile of CA-PCPP-PDADMAC-CA block co-polymer was performed at different temperatures and different buffer solutions in Figure. 5.10. CA grafted polymer exhibits a pH and temperature-based swelling nature. At body temperature, the di-block polymer has reached a maximum of 83 % of swelling at pH 5.0. In acidic pH, the number of ionizable groups in the polymeric system significantly increased due to a rise in the temperature. So, the swelling of nanomaterials was high at 37 °C compared to 25 °C (72 % swelling). The swelling of polymeric nanomaterial was low at neutral pH, it forms 67 % of swelling at 37 °C and 58 % swelling at 25 °C. Polymeric -CH layer swells and becomes more viscous in pH 5.0 and becomes elastic in pH 7.0. There is no noticeable change in the swelling capacity of polymers in neutral pH of low temperature. This characterizes the responsive behavior of di-block polymers under various physiological conditions, so that has been selected to be used in drug delivery applications. The swelling property of the polymer was engaged by several parameters including the amount of the cross-linking and ionic character, a high degree of the cross-linking decreases the swelling property of the polymer. CA-PCPP-PDADMAC-CA from the high amount of cross-linking at neutral pH and in acidic pH, there was an increase in the ion concentration which raised the swelling ability of polymers [153,154].

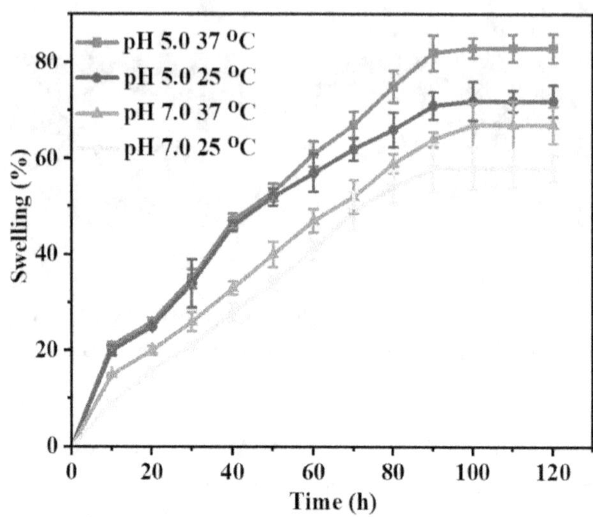

Figure. 5.10. Swelling behavior CA-PCPP-PDADMAC-CA polymer in pH 5.0 and 7.0 at 25 and 37 °C.

5.6.4. Encapsulation and *in vitro* drug release

Encapsulation of PTX in polymeric nanomaterial is a key parameter to evaluate the therapeutic efficacy of nanocarriers *in vitro* (Figure. 5.11). By changing the concentrations of polymer alters the drug encapsulation effect. Polymer concentration is the important parameters for self-assembly to produce the desired vesicles. The drug could uniformly disperse on the nanomaterial to fully contact the functional group [155].

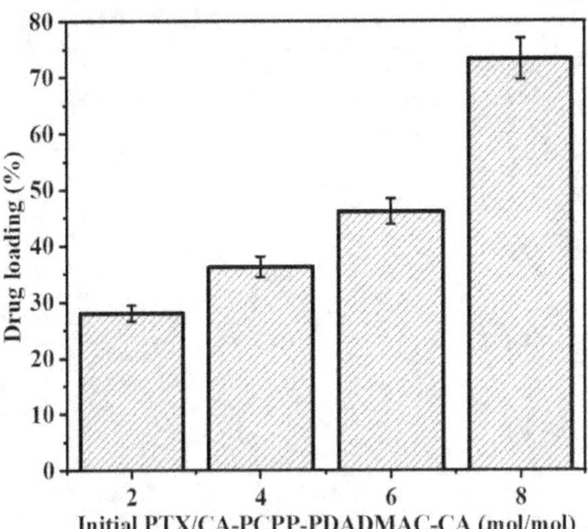

Figure. 5.11. Different concentrations of PTX encapsulation in CA-PCPP-PDADMAC-CA.

Figure. 5.12, reveals the cumulative PTX release profiles of polymeric nanomaterial was performed by dialysis method. The CA grafted polymer exhibits drug release based on pH conditions. Initially, polymer burst release the PTX followed by release drug in a controlled manner. Polymeric nanomaterials allowed 90 % release of PTX within 6 days at pH 5.0 and 43 % release at pH 7.0. The drug release was decreased as the pH of the media increased. This was partially responsible for the differences in drug release in alternate pH conditions. Because the acidic condition is accelerated by the protonation and thereby, the drug PTX has been released easily from the hybrid polymeric carrier. Higher bioavialbility of drug in tumor cell was due to the acidic environement [156].

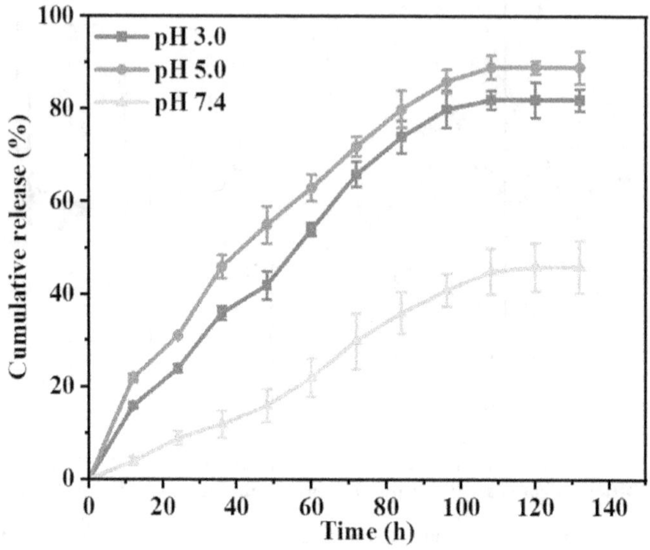

Figure.5.12. The cumulative release profile of PTX loaded CA-PCPP-PDADMAC-CA polymeric nanomaterial in pH 3.0, 5.0, and 7.4. at 37 °C.

5.6.5. Cytotoxicity assay

The *in vitro* cytotoxicity of PTX loaded CA-PCPP-PDADMAC-CA polymeric nanomaterial was evaluated against MCF-7 cells by MTT assay. CA-PCPP-PDADMAC-CA polymeric nanomaterial exhibited no cytotoxicity even at 200 µg/mL, representing the high biocompatible nature of cholate grafted polymer (Figure.5.13. A). Cells were shows dose-dependent toxicity after treatment with PTX loaded polymeric nanomaterial. At pH 5.0, free PTX and PTX loaded polymeric nanomaterial shows cytotoxicity at low concentration, and polymeric nanomaterial was more effective than the free drug. This effect was due to the presence of cholate in the polymer and the bioavailability nature of the nanocarrier. In Figure.5.13. B, PTX polymeric nanomaterial show IC_{50} of 5 µg/mL (pH 5.0) and 10 µg/mL (pH 7.0). This varied cytotoxic effect of polymeric nanomaterial was due to the pH-responsive drug release system. In pH 5.0, CA-PCPP-PDADMAC-CA activated by the protonation and the maximum amount of drug released from nanomaterial and there was very less amount of drug release at pH 7.0. This leads to efficient suppression of MCF-7 cells at pH 5.0 [157].

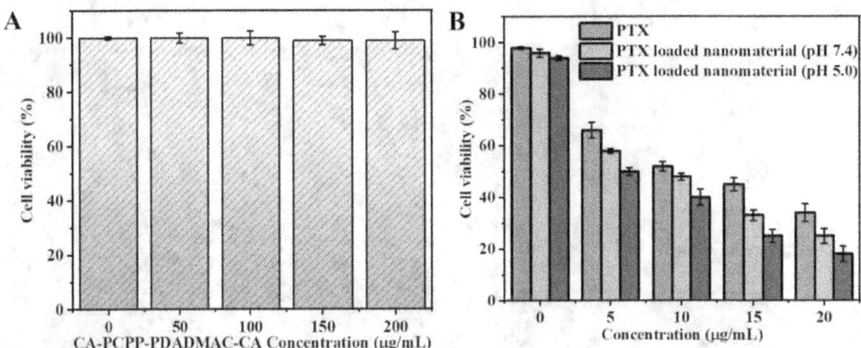

Figure. 5.13. (A) Cytotoxicity of CA-PCPP-PDADMAC-CA with different concentrations on MCF-7 cells after 24 h incubation (B) MCF-7 cell viability after 24 h incubation of PTX and PTX loaded CA-PCPP-PDADMAC-CA with different dosage at pH 7.0, 5.0.

5.6.6. γH2AX assay

The CA functionalized polymer was internalized into cells and release PTX drug in the cytoplasm of MCF-7 cells leads to DNA damage were confirmed by the γH2AX assay (Figure. 5.14). Here, we evaluate the γH2AX induction after exposure to PTX loaded polymeric nanomaterial on MCF-7 cells. The control group treated with polymeric nanomaterial without drug did not show increased γH2AX expression due to the absence of PTX. Increased γH2AX foci were enumerated in MCF-7 cells by time depending on manner. And also, substantiating the similarity of nanomaterial sensitivities between the different time intervals, MCF-7 cells treatment with CA-PCPP-PDADMAC-CA nanomaterial of 12 and 24 h were compared for γ-H2AX kinetics. Polymeric nanomaterial after treatment with MCF-7 cells shows 1.8-fold γH2AX density increases for cells in each treatment. There was high induction of γH2AX expression in 24 h treatment compared to 12 h treatment. Exposure of polymeric nanomaterial for 24 h makes chromosomal double-strand breaks associated with apoptosis and chromatin compaction in early mitosis. It inhibits the topoisomerase activity and arrests the cells at S phase or translocates [158]. This strong signal is due to the presence of CA on the surface of polymeric nanomaterial and it is an ideal option for targeted drug delivery. And CA derivatives can also inhibit the growth of the MCF-7 breast carcinoma cells by DNA damage [159].

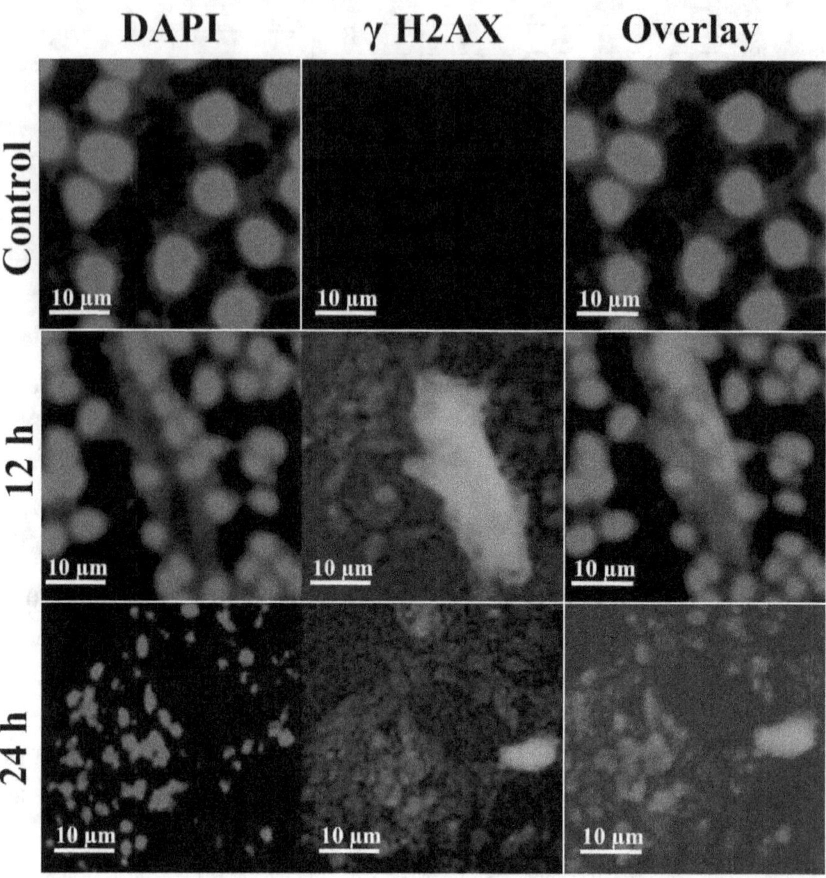

Figure. 5.14. γH2AX assay of cells treated with CA-PCPP-PDADMAC-CA for 12 h and 24 h.

5.6.7. Comet assay

PTX loaded polymeric nanomaterial induced DNA damage in MCF-7 cells were analyzed using comet assay by tail length and percentage of DNA fragments in comet tails. There was no damage detectable on untreated cells and DNA damage in treated cells was evident by the comet tail (Figure.5.15. A). Therefore, we compared the levels of DNA damage in MCF-7 cells after treatment with PTX loaded polymeric nanomaterial at different time intervals. Polymeric nanomaterial exhibited a robust potential of PTX on DNA damage in a time-dependent manner. In both 12 and 24 h treatments represents in Figure.5.15. B, C, over 50 % of total DNA was fragmented and located outside the nucleus versus the control group. The tail length and the percentage of DNA in the comet tail increase by 1.86 to 2.35-fold compared to 12 h treated groups, respectively. CA functionalized polymeric nanomaterial targeted the tumor site and the EPR effect of nanomaterial degrades cancer DNA. Nanomaterial increases free radical leads to chromosome damage and instability of the genome. After threshold DNA damage has occurred, apoptosis-activating enzymes cascade may be initiated at different exposure levels for different cell phenotypes, leading to cell death [160].

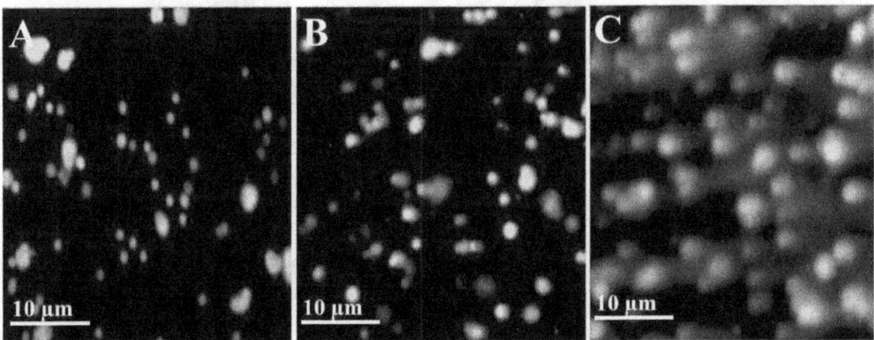

Figure. 5.15. Comet assay images of cells treated by CA-PCPP-PDADMAC-CA (A) Control (B) 12 h (C) 24 h.

5.6.8. Western blotting

The effect of PTX loaded polymeric nanomaterial on apoptosis mRNA expression and protein levels p53, Bax, and Bcl-2. These proteins are related to the mitochondrial apoptotic pathway. The nanomaterial significantly alters the protein levels of Bax, Bcl-2, and p53 was compared with control (Figure.5.16. A). In Figure.5.16. B, C, data confirms increased levels of Bax and decreased Bcl-2 proteins and it clearly shows the time-dependent changes in the protein expression. CA-functionalized polymeric nanomaterial was enhanced the anticancer activities and causes apoptosis through a decline in mitochondrial function, activation of p53 and pro-apoptotic Bax, and decreased expression of Bcl-2. Caspases play important role in the activation of cell apoptosis and the change of the Bax/Bcl- 2 ratio could result in significant activation of caspases, and lead to cell death. CA present in polymeric nanomaterial enhances the EPR effect and also activates the FXR which down-regulates the expression of breast cancer cell proliferation genes [161].

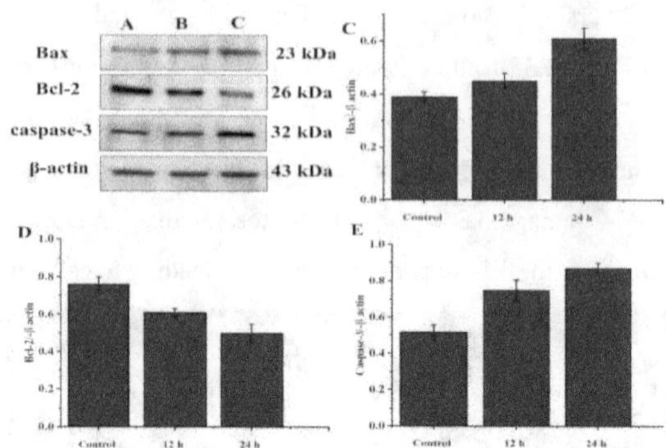

Figure. 5.16. Expression of Bax, Bcl-2, caspase-3, and β-actin in MCF-7 cells after treatment with IC$_{50}$ concentration of PTX loaded CA-PCPP-PDADMAC-CA (A) Control (B) 12 h (C) 24 h was analyzed western blot, respectively. (D) The relative protein expression of Bcl-2 which compared with that in the control group (E) The relative protein expression of Bax which compared with that in the control group (F) The relative protein expression of caspase-3 which compared with that in the control group. The same blots were stripped and re-probed with a β-actin antibody to show equal protein loading.

5.7. Conclusion

To summarize, the development of a CA functionalized polymeric nanomaterial targeted drug delivery system to improve the therapeutic efficiency of PTX without systemic toxicity. In this study, CA which has carboxyl groups conjugated into PCPP using ethylenediamine and sodium cholate forms ionic interaction of PDADMAC. The resulting polymers were used for the formation of CA-PCPP-PDADMAC-CA polymeric nanomaterial with better dispersibility and high loading capacity of hydrophobic drug PTX. Polymeric nanomaterial exhibit temperature and pH-based responsiveness during swelling and drug release. It helps polymeric nanomaterial could successfully deliver the drug to the targeted site. And also, CA functionalized nanomaterial recognizes FXR receptors containing breast cancer cells and is selectively endocytosed into the cancer cells. Owing to their acidic pH, the nanomaterial disassembles and is subsequently releases PTX towards the cytoplasm and perinuclear region to reduce the mitochondrial transmembrane potential and mediate the apoptotic cell death pathway. Targeting breast cancer cells by CA moiety was confirmed by efficient cellular uptake, intracellular release, and DNA damage of cells. Hence, the stimuli sensitive and FXR receptor-targeted CA-PCPP-PDADMAC-CA co-polymeric nanomaterial are considered to be excellent candidates for the co-delivery of the PTX at specific breast cancer sites to attain an improvement in chemotherapeutic treatment.

6.1. Introduction

Nano-drug delivery has focused more attention on breast cancer a major life-threatening illness for women. However, these nanocarriers encounter various barriers during the route of injection like non-specific uptake, mucosal barriers, extravasation in non-target sites [162-164]. Therefore, alternative therapeutic approaches are needed to develop, which overcome the problems. Transdermal drug delivery is a good strategy; it bypasses primary metabolism, non-invasive, and maintains a constant drug level. It also has an advantage like lower toxicity, higher therapeutic efficacy, and lessen the periodic chemotherapeutic administrations [165,166]. For transdermal drug delivery systems, nanofibers carriers drug depots like liposomes, emulsion, microsphere, micelles. So, incorporation of ligand coupled drug depots into nanofibers material, to form a core-shell system. It fabricates through coaxial electrospinning, consist of two capillaries placed together coaxially [167]. It contains shell solution (spinnable material) injected through the outer needle and core solution (drug depot) through the inner needle to produce nanofibers with hollow space [168,169]. Different types of strategies were reported for controlled drug release of the core-shell nanofibers system. The core-shell structure is composed of nano, and micro-size ranged fibers with pores. The drug release controlled in core-shell nanofibers was due to the outer shell region prevents the rupture and protects the active components.

6.2. Previous work

A partially electro-spinnable shell solution was used for the fabrication of drug-loaded fibers. It exhibits a sustained release of drugs into cancer tissue. Targeted drug delivery was achieved by encapsulation of drug-loaded active targeting nanomaterial into the poly (vinyl alcohol) polymers. Core-shell electrospinning of poly (L-lactic acid-co-ε-caprolactone) exhibited controlled release of PTX for 60 days and effectively kills the HeLa cells [170,171]. Doxorubicin-loaded porous silica nanoparticles coupled with PLA nanofibers exhibited high stability in structure and controlled drug release [172].

6.3. Novelty

The mucilage obtained from *Plantago ovata*, could be considered as the primary precursors for fabrication of core-shell nanofibers. *P. ovata* (dietary fibers) mucilage was suitable for the electrospinning process because of charge density. It contains more hydroxyl and carboxyl groups. US-FDA approved, biodegradable psyllium husk mucilage (PHM) for food packaging. It has strong mechanical strength and the ability to encapsulate the drug-loaded nanomaterial into PHM [173,174]. So, it is a promising source to fabricate core-shell nanofibers.

6.4. Objective

The main aim of the study is to fabricate ligand coupled nanomaterial into biodegradable nanofibers, for pH-responsive and targeted cancer drug delivery. Initially, CA conjugated to the poly (bis (carboxyphenoxy) phosphazene) (PCPP) polymeric nanomaterial that was used to encapsulate PTX. CA ligand specifically helps in nanomaterials internalization into the breast cancer cells due to a strong affinity for FXR. Further, the nanomaterial encapsulates into the core-shell PHM nanofibers. So, the resultant nanofibers are placed subcutaneously near the solid tumor region. Nanofibers matrix degradation causes controlled release of active targeting nanomaterial which increases the concentration of drug in the tumor area for a longer time. We characterized the pH-responsive *in vitro* and *ex vivo* drug release ability of nanofibers at different conditions. The effect of the active targeting nanomaterial loaded nanofibers on the biocompatibility, viability, cytotoxicity of cancer cells and noncancerous cells is evaluated. Nanofibers matrix and active targeting nanomaterial system combination endow novel *in situ* drug delivery with high therapeutic efficacy, low toxicity, and low treatment costs.

6.5. Experimental techniques
6.5.1. Preparation of PTX loaded PCPP-CA nanomaterial

Briefly, 100 mg of PCPP-CA conjugate was dissolved in 10 mL water and dispersed by a probe sonicator for 5 min. PTX in 5 mL chloroform (30 mg/mL) was added to the PCPP-CA conjugate solution, and vigorously stirred for 20 min. The mixture solution was ultrasonicated for 30 min and then dialyzed against deionized water for 24 h. Finally, nanomaterials were lyophilized and stored at -20 °C for further use.

6.5.2. PHM extraction and electrospinning solution preparation

Psyllium husk soaked in water at a 50:1 ratio under constant stirring overnight at 45 °C. Mucilage from the swollen husk filtered through a muslin cloth and centrifuged at 7000 rpm for 15 min. Then freeze-dried to obtain as a powder and stored in a desiccator. PHM powders blended in acetic acid at different ratios (20, 25, 30 % w/V) with 2 % glycerol for electrospinning. The sample flow rates were fixed at 0.15 mL/h and electric potential was maintained at 20 kV consistently. Nanofibers were collected by a grounded plate covered by aluminum foil.

6.5.3. Electrospinning of PTX@PCPP-CA in PHM nanofibers

To incorporate PTX loaded nanomaterial in the mucilage nanofibers matrix, the coaxial needle method was used. Coaxial electrospinning prepared by inner needle (inner diameter 0.8 mm) and outer needle (inner diameter 0.35 mm) positioned coaxially. The outer needle consists of shell solution (PHM), and the inner needle carries a core solution (nanomaterials) to form core-shell nanofibers. Homogenous dispersion of PHM (20 %) and PTX loaded PCPP-CA nanomaterial (10 %) (w/v) in the ratio of 1:3, 2:3 and 3:3 (w/w) ratio used for electrospinning. The inner and outer fluids flow rates were fixed at 0.15 mL/h and electric potential was maintained at 20 kV consistently. Nanofibers collected by a grounded plate covered by aluminum foil and the distance between the needles, and the plate was 15 cm. The process conducted out at 25 °C and 45 % relative humidity.

6.5.4. *Ex vivo* skin permeation and drug release studies

Briefly, 500 µm thickness of pork skin (provided by a local slaughterhouse-after 30 min slaughtering) collected in Krebs buffer and extraneous subcutaneous fat removed from the dorsal surface. Skin membrane mounted between the donor and receptor compartments of vertical Franz type diffusion cells and epithelial side facing the donor compartment with an active permeation area of 1.7 cm^2. Electrospun nanofibers of 0.2 mg were applied to the epithelial surface in the donor compartment. Receptor compartment filled with various buffer pH to identify pH-based skin permeation. A buffer of 2 mL was withdrawn periodically from the receiver end, to evaluate the amount of drug released and replaced with an equal amount of fresh buffer. The sample solutions were examined by a UV-vis spectrometer at 229 nm. After skin permeation studies, nanofibers were removed from Franz-type diffusion cells, samples collected onto a copper grid covered with a carbon film, and nanofibers visualized by TEM imaging.

6.5.5. Confocal microscopy analysis

Confocal microscopy images of the skin were recorded after treatment with FITC tagged nanofibers. The permeation assay performed using Franz-type diffusion cells experimental conditions described above. After 8 h, the skin samples were cleaned and sectioned using a microtome. The skin is coated on a glass slide and images are captured through the coverslip side. The full epidermis was scanned at different increments through the Z-axis of the confocal microscope (Carl zeiss-Model : LSM 700).

6.5.6. Cytotoxicity assays

The cytotoxicity studies were performed in two methods, In the first method, MCF-7 cells were seeded into the nanofibers (PCPP-CA-PHM, PTX, PTX-PCPP-CA-PHM) surface at a density of 4×10^4 cells. In another method, nanofibers (PCPP-CA-PHM, PTX, PTX-PCPP-CA-PHM) are fixed between the two glass rings and are completely immersed into the cells attached plate. Both were maintained in a humidified environment at 37 °C with 5 % CO_2 for 8 days. The only difference in nanofibers fixed in the glass ring method is samples introduced after cell attachment in the plate. The medium is replaced every 2 days in both methods. Nanofiber's cytotoxicity ability was measured in different time intervals (days 1, 4, and 8). MTT solution (50 µL) was added to each well, kept for 4 h incubation, and after the addition of 100 µL DMSO per well, OD was measured at 595 nm.

6.5.7. Cell morphology

PTX-PCPP-CA-PHM nanofibers coated on the glass coverslip placed on the culture medium contained a 6-well plate. Cells were seeded in the plate with a density of 4×10^4 cells/well and the plate maintained for 24 h, then the cells in the coverslip were washed twice with PBS and fixation using 2 % glutaraldehyde for 10 min followed by 20 %, 40 %, 60 % and 80 % ethanol gradient. The dried sample sputter-coated with gold to examine the cell membrane integrity under SEM.

6.6. Result and discussion
6.6.1. Characterization of the PCPP-CA nanomaterial in nanofibers matrix

The electrospinning technique has developed to fabricate the PTX loaded nanofibers in a core-shell structure. In this PTX, encapsulated nanomaterial was surrounded by PHM. The coaxial method helps in the protection of the nanomaterials, site-directed delivery, high encapsulation of nanomaterial into the fiber [175]. CA has a high affinity towards the FXR receptors contained in breast cancer cells. The carboxyl group of CA and PCPP linked with ethylenediamine and PTX solution introduced into the PCPP-CA to form polymeric nanomaterial.

(i) PTX@PCPP-CA nanomaterial SEM analysis

The CA functionalized PCPP nanomaterial size 110 nm (Figure. 6.1. A) and it was spherical shape which confirmed by SEM (Figure.6. 1. B). The active CA functionalized PCPP nanomaterial delivers the drug to breast cancer.

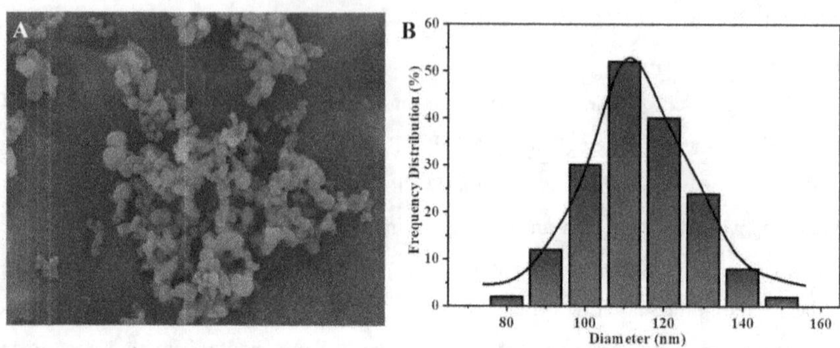

Figure. 6.1. (A) SEM image representing morphology of PTX loaded PCPP-CA (B) DLS curve of PTX loaded PCPP-CA.

(ii) PHM nanofibers SEM analysis

In nanofibers preparation, PHM concentration and solvent ratio essential for making the desired morphology. Homogenously PHM dispersed in the acetic acid solution for the fabrication of electrospinning. In both concentration of 20 % and 25 % PHM form uniform morphology which represents in Figure.6.2. A, B. It was tedious to form nanofibers at high concentrations (30 %), but at a lower concentration, it forms continuous nanofibers. It may be due to the slow evaporation rate of acetic acid at high concentrations. PHM nanofiber formed in size of 220 nm (20 %), 240 nm (25 %) represented in Figure. 6.2. C, D. PHM has several advantages like good electrospinning properties, high compatibility with nanomaterials, high viscosity, and conductivity [176]. The nanomaterials and PHM component ratio 1:3 (w/w) were selected due to the more stability.

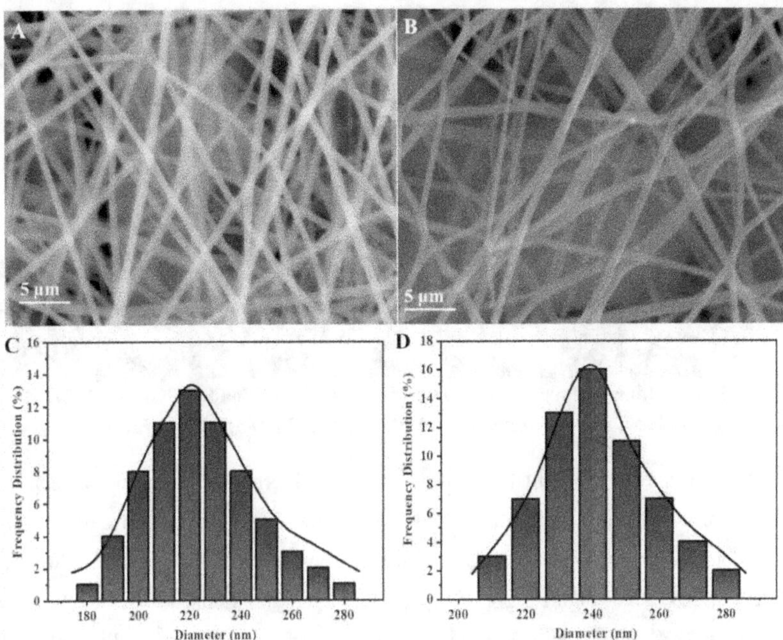

Figure. 6.2. (A) SEM image of 20 % of PHM nanofibers (B) 25 % of PHM nanofibers morphology represents in SEM. Size range of (C) 20 % of PHM nanofibers (D) 25 % PHM nanofibers.

(iii) PTX loaded PCPP-CA-PHM nanofibers SEM analysis

As illustrated in Figure. 6.3. A, B, the component ratio of 2:3, 3:3 not exhibited uniform morphology. The changing composition of the polymer solution has a strong effect on the nanofiber's morphology and an adequate adjustment of mucilage. In essence, the nanomaterial ratio can lead to nanofibers showing homogeneous shapes and a narrow width distribution, to warranty a higher surface area [177].

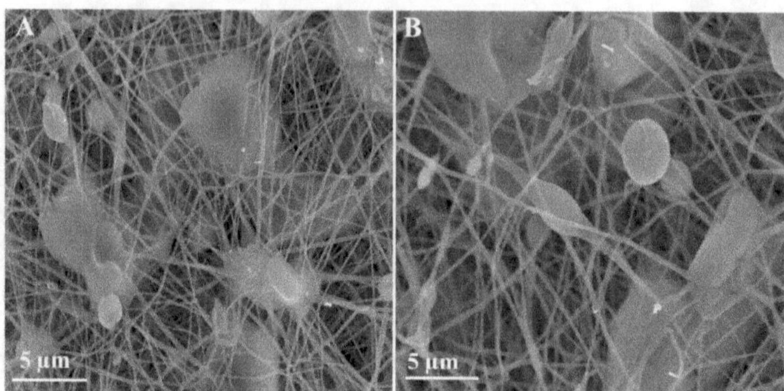

Figure. 6.3. SEM image representing size and shape of PTX loaded PCPP-CA-PHM nanofibers with component ratio of nanomaterial and PHM (A) 2:3 (B) 3:3 (w/w) ratio.

PTX loaded PCPP-CA-PHM nanofibers (1:3) were formed successfully by electrospinning and the structural morphology of nanofibers denoted by SEM is shown in Figure. 6.4. A-D. Nanofibers were randomly distributed to form the continuous fibrous structure and smooth surfaces with uniform diameters of 239±2 nm (Figure. 6.5). It appears that after the introduction of PCPP-CA nanomaterial into the inner core, there is no change in the diameter of the nanofibers. Additionally, in some spots the diameter of the nanofibers became ununiform, and "slight enlargements" were found in the PCPP-CA-PHM nanofibers suggesting that this nanomaterial was successfully encapsulated into the core regions of the nanofibers.

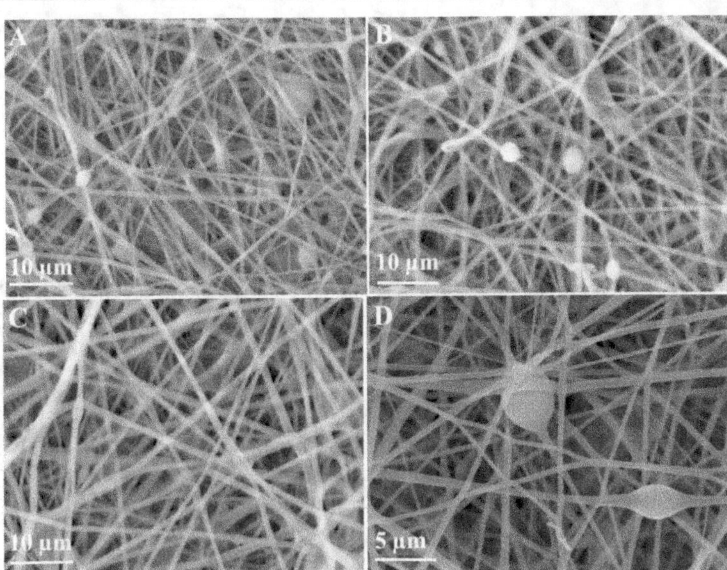

Figure. 6.4. SEM image representing the shape of 1:3 (w/w) ratio PTX-PCPP-CA-PHM nanofibers on the different region (A) 10 μm (B) 10 μm (C) 10 μm (D) 5 μm.

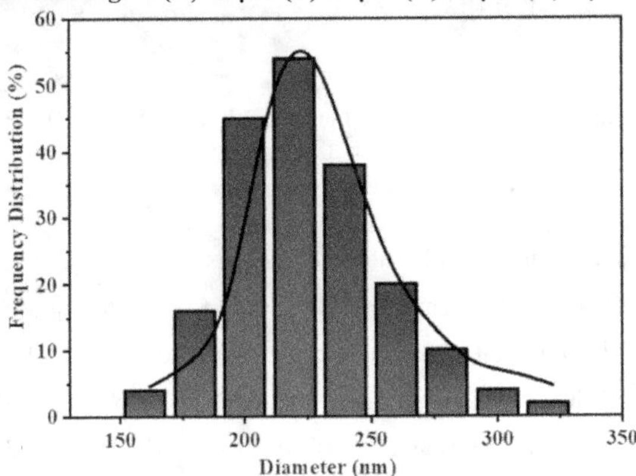

Figure. 6.5. Diameter of PTX-PCPP-CA-PHM nanofibers by DLS curve.

(iv) AFM analysis

AFM images confirmed the morphology and topography of the nanofibers in 2D image (Figure. 6.6. A, C). The nanofibers showed a cylindrical, continuous, and 3D-network structure similar to that of the nanofibers SEM image (Figure. 6.6. B,D). Nanofibers size measured with AFM was higher than those obtained by SEM, because of bump effect nearby fibers. The topography of PCPP-CA-PHM nanofibers was smooth and without agglomeration indicating no accumulation of nanomaterials at the surface of the nanofibers. A high proportion of nanomaterials are properly encapsulated and distributed inside the nanofibers.

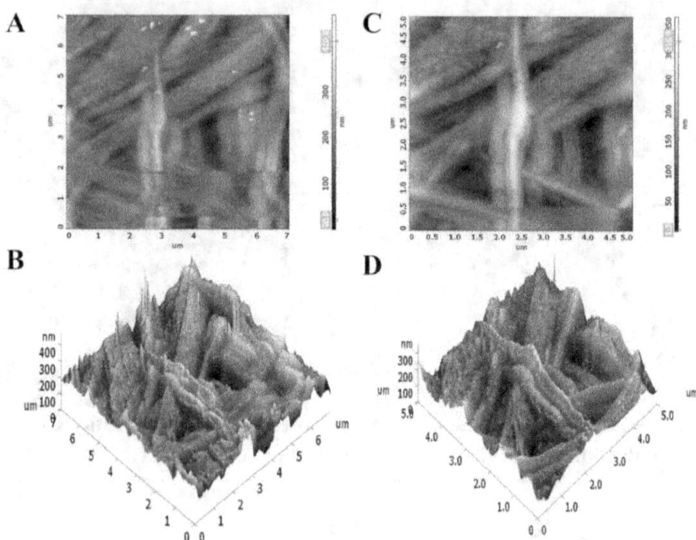

Figure. 6.6. AFM image representing surface morphology of PTX-PCPP-CA-PHM nanofibers (A, C) 2D image (B, D) 3D image.

(v) FTIR analysis

The FTIR spectrum of PCPP shows characteristic absorption bands like 672 cm^{-1} (C-H vibration), 1317 cm^{-1} (P=N), 2727 cm^{-1} (carboxylic acid O-H stretching), 3027 cm^{-1} (aromatic C-H) (Figure. 6.7. A). A sharp peak at 1704 cm^{-1} attributes the cholate -COOH stretching and 2831 cm^{-1}, 1296 cm^{-1}, 1402 cm^{-1} bands denote the following functional groups like C-H stretching vibration, hydroxyl, skeletal C-C vibration (Figure. 6.7. B). PCPP-CA exhibited the adsorption band 1686 cm^{-1} denotes the C=O stretching groups and 782 cm^{-1} represents C=C stretching (Figure. 6.7. C). Hence, from this observation, it can be interpreted that a decrease in carboxyl -OH stretching due to maximum interaction of PCPP with CA. The PHM forms OH stretching bands at 3386 cm^{-1}, C-H stretching vibration at 2934 cm^{-1}, -CH$_3$ symmetric at 1403 cm^{-1}, C-C and C-O bonds at 1036 cm^{-1} (Figure. 6.7. D). Nanofibers consist of adsorption bands 1352 cm^{-1}, 3426 cm^{-1}, 1807 cm^{-1} represents the functional groups such as C-H group, hydrogen-bonded O-H stretching, -CH$_2$ bands (Figure. 6.7. E). Functional group like carboxyl, hydroxyl, phosphorous groups presence in PCPP nanofibers denotes the formation of PCPP-CA-PHM nanofibers formation. PHM and PCPP-CA-PHM nanofibers band shifts are similar, which may be due to the strong interactions and higher degree of cross-linking among mucilage and polymer.

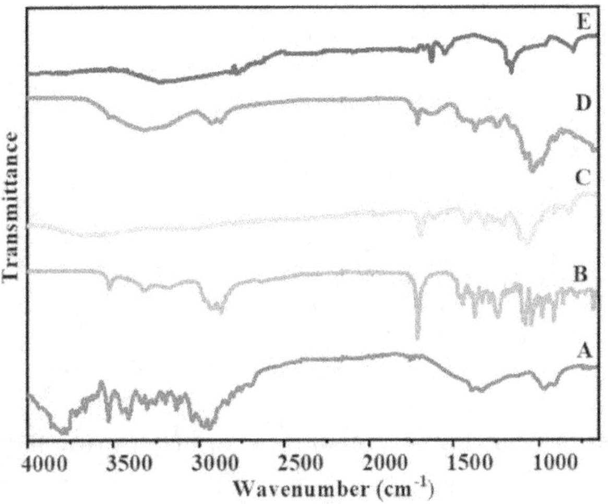

Figure. 6.7. FTIR spectroscopy of (A) PCPP (B) CA (C) PCPP-CA (D) PHM (E) PCPP-CA-PHM nanofibers.

(vi) XRD pattern

To investigate the changes in the PCPP, CA, PHM microstructure, and nanofibers, XRD diagrams. PCPP had typical diffraction peaks at 2Θ - 25.7°, 26.6°, 27.3°, 31.8°, 23.9°, and there were smaller peaks in the range 20° to 25° (Figure. 6.8. A). XRD pattern of PCPP exhibits more spike peaks implies the crystalline nature of the polymer. The diffraction curve of CA showed five typical crystal peaks at 2Θ equals 21.6°, 23.4°, 24.1°, 26.4°, and 34.2° (Figure. 6.8. B). PCPP, CA functionalization with ethylenediamine to form PCPP-CA complex shows PCPP typical crystal peak with minimum intensity at 2Θ - 26.2°, 26.9°, 31.8° along with CA peak 21.6°, 23.4°, 24.8° which indicates the CA successfully embedded on the polymer matrix (Figure. 6.8. C). PHM presented crystalline characteristics and exhibits sharp peaks (Figure. 6.8. D). The changes in the diffraction pattern indicate the formation of electrospun nanofibers. It shows a diffraction peak similar to PCPP-CA, PHM, and partial peaks disappeared (Figure. 6.8. E). The intensity and the peak location changed obviously due to the electrospinning process [178]. Additionally, the crystallization rate of nanofibers was reduced due to the polymer structure, molecular weight, and also polymer-solvent relations [179].

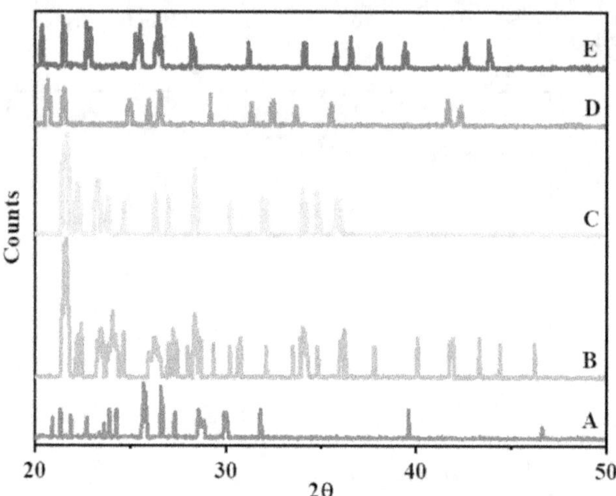

Figure. 6.8. XRD pattern of (A) PCPP (B) CA (C) PCPP-CA (D) PHM (E) PCPP-CA-PHM nanofibers.

(vii) TGA studies

TGA was performed to determine the thermal stability of PCPP-CA-PHM nanofibers (Figure. 6.9). The 50 % mass loss of PCPP polymer in the range of 330 °C was due to its decomposition of the polymer. Meanwhile, PCPP-CA showed a slightly different behavior than PCPP, it shows a high 10 % mass loss when temperature increased above 250 °C. But contrary PCPP-CA-PHM nanofibers mat thermal transition appears at a higher value than PCPP, PCPP-CA. Nanofibers display mass loss started at temperature 320 °C, and this transition is associated with degradation of PHM polymeric backbone. The nanofibers were more stable than the polymers alone, and the inclusion of nanomaterials in the nanofibers did not affect the thermal profile.

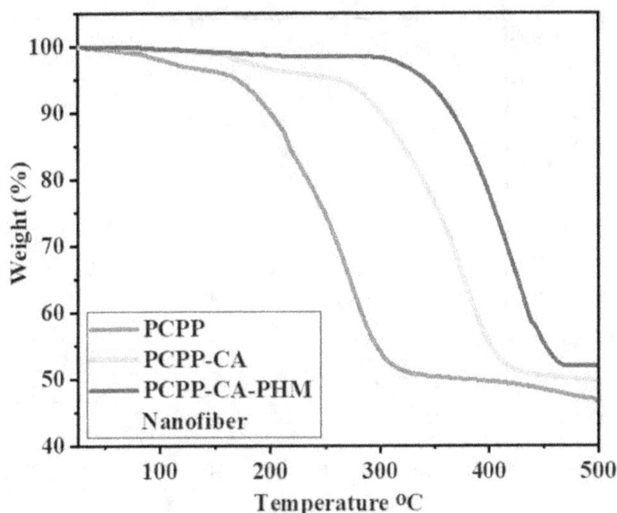

Figure. 6.9. TGA curve analysis of PCPP, PCPP-CA, PCPP-CA-PHM nanofibers.

6.6.2. Swelling studies

The swelling and degradation properties of the PCPP-CA-PHM nanofibers were subsequently analyzed via different pH at 37 °C, mimicking physiological conditions (Figure. 6.10). The degree of swelling of nanofibers plays an important role in the loading and release behavior of PTX. The degree of swelling of nanofibers in pH 7.0 is 20 %, 140 %, 110 %, and 75 % and pH 5.0 swelling percentage was 50 %, 175 %, 155 % and 120 % for time interval 40, 80, 120 and 160 h respectively. The swelling occurred rapidly by maximum water uptake achieved 175 % at 80 h and followed by deswelling over than next 60 h. Nanofibers degree of swelling decreased due to the PCPP-CA nanomaterial that can make the interaction with PHM polymeric chain that restricts the swelling as a consequence of nanofibers are not able to attain maximum swelling [180]. The rapid and higher magnitude swelling on acidic pH facilitated the nanofibrous network structure is of potential interest in transdermal drug delivery applications. At pH 5.0, the degradation of the PHM cross-link was in higher condition results in increased swelling.

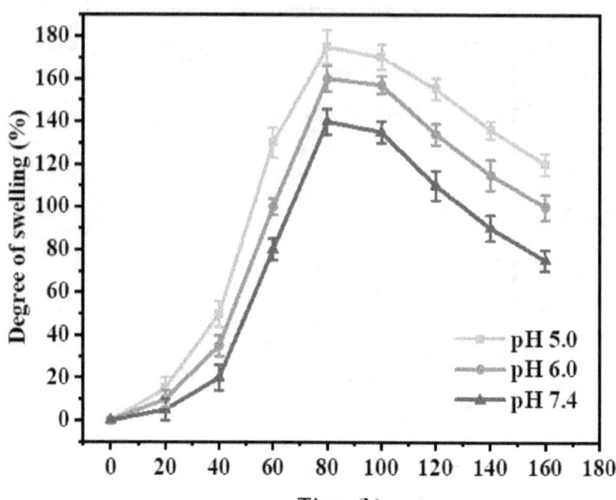

Figure. 6.10. Swelling behavior of PTX-PCPP-CA-PHM nanofibers at pH 5.0, 6.0, 7.4.

6.6.3. Weight loss analysis

Weight loss analysis (Figure.6.11) was used to assess the degradation profile of PCPP-CA-PHM nanofibers. It revealed an initial weight loss of around 15-18 % on day 5 and a random 7-10 % weight loss of nanofibers for every 10 h. Changes in pH values were monitored over the experimental period showing higher weight loss over-acidic pH. Weight loss is attributed to the gradual degradation of the cross-linked outer shell, the dissolution of the inner PCPP-CA regions, and the release of more nanomaterials.

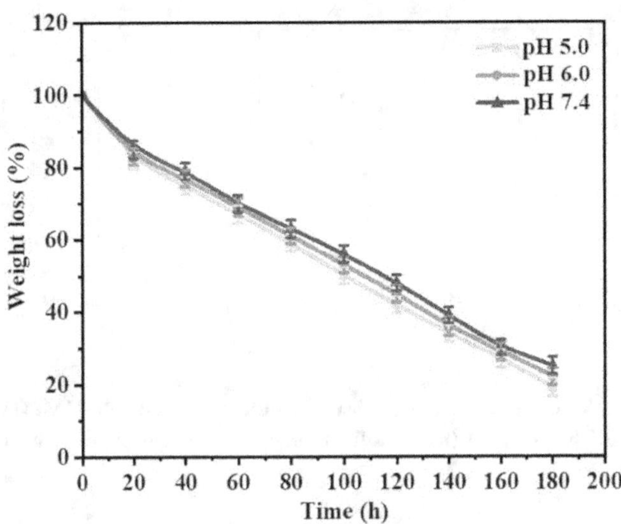

Figure. 6.11. Weight loss ability of PTX-PCPP-CA-PHM nanofibers at pH 5.0, 6.0, 7.4.

6.6.4. *In vitro* drug release

Two different strategies were used for *in vitro* drug release, the dialysis bag method (Figure. 6.12. B) and the Franz diffusion cell apparatus (Figure. 6.13. B). In the dialysis bag method, nanofibers initially release PTX rapidly at 100 h, followed by a slow release for 180 h in different pH (Figure. 6.12. A). PTX was released from the nanofibers outside the dialysis bag to burst release in the 36 % in 20 h at pH 5.0. It had a cumulative release reaching approximately 86 % in 140 h at pH 5.0. In pH 7.4, the amount of PTX released from the PCPP-CA-PHM nanofibers system, while almost 74 % at 200 h and 80 % at pH 6.0. These results indicate that the nanomaterials before the dispersion in the nanofibers were able to provide a slower and sustained delivery of the drug, which is desirable for a controlled release system [181]. The amount of released drug was increasing in order in pH 5.0 compared to pH 6.0 and 7.4. PCPP-CA-PHM nanofibers have excellent transdermal drug delivery properties because the skin pH (4.0-5.5) was acidic. Higher drug release at acidic pH was due to the erosion of the PHM outer shell, drug diffusion, and polymer degradation. Nanofibers matrix composed of PHM polymers, adsorb H^+ ions and repulsion between H^+ ions promoting swelling and the burst release of the PTX. Subsequently, the surface is eroded, followed by continuous and controlled diffusion of the drug [182].

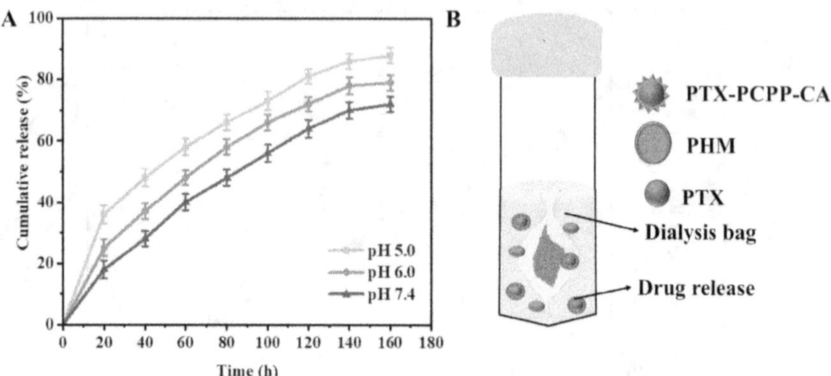

Figure. 6.12. (A) Cumulative PTX release profile of PTX-PCPP-CA-PHM nanofibers from dialysis bag pH 5.0, 6.0, 7.4 **(B)** Nanofibers packed in a dialysis bag and immersed in buffer solution.

6.6.5. *Ex vivo* skin permeation study

The skin permeation study was performed to evaluate the effect of the nanofibers on permeation through skin layers and this was examined through Franz diffusion cell apparatus (Figure. 6.13. A) under different pH conditions. It predicts in advance how the nanomaterial will behave *in vivo*. The cumulative amount of PTX permeating through the pork skin was 28 %, 20 %, 10 % over 20 h at pH 5.0, 6.0, 7.4. This evidence nanomaterial permeation across the skin and the internalization inside the skin was low at pH 7.4, and the high amounts of drug deposition were detected on the skin surface at pH 5.0. There was a steady increase in nanomaterials permeation of 70 %, 62 %, and 50 % at 120 h (pH 5.0, 6.0, 7.4). Skin retention at acidic pH was higher than neutral pH, probably due to the maximum degradation at pH 5.0, and PHM managed to increase skin retention of nanomaterials [183,184].

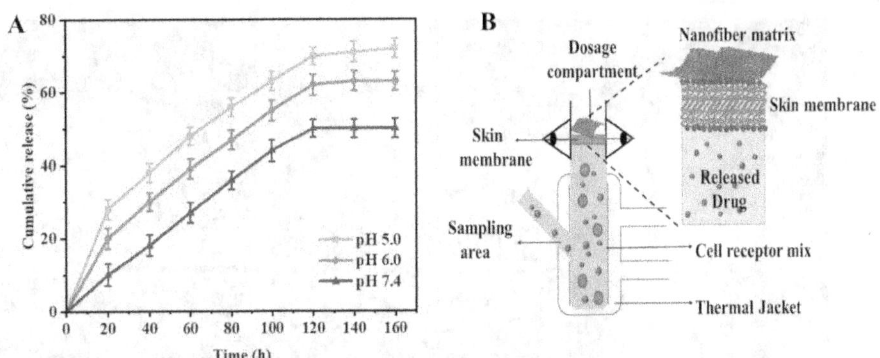

Figure. 6.13. (A) *Ex vivo* **skin permeation of PTX from nanofibers mat through Franz type diffusion cells pH 5.0, 6.0, 7.4 (B) Franz type diffusion cells apparatus based drug release through the skin membrane.**

6.6.6. Confocal microscopical analysis

To confirm the skin permeation ability of PCPP-CA-PHM nanofibers, confocal microscopic studies were conducted. Skin samples were treated with fluorescein nanomaterial-loaded nanofibers using Franz-type diffusion cells under different pH for 8 h. In Figure.6.14. A,D, indicate the localization of green fluorescein on the outer layer of the skin samples (pH 5.0) treated with nanofibers emits high fluorescence compared with the nanofibers treatment with another pH 6.0, 7.4 (Figure.6.14. B, E, C, F). There is a pH-based difference in the internalization of fluorescein dye at the outer layer of the skin due to the maximum release of dye at lower pH. The skin permeation profiles of the nanofibers are in agreement with the results of the release studies. Nanofibers increase the skin's internalization ability by segregating the drug-loaded nanomaterial into the skin. It enhances the permeation of drug through stratum corneum and nanoparticles, passes through pores close to hair follicles [185]. Additionally, breast skin attached with an internal lymphatic circulation and availability of retro mammary fatty envelope, serve as a drug reservoir for long-time drug distribution to breast cancer [186]. So, *in-situ* transdermal delivery is a suitable technique for breast cancer therapy.

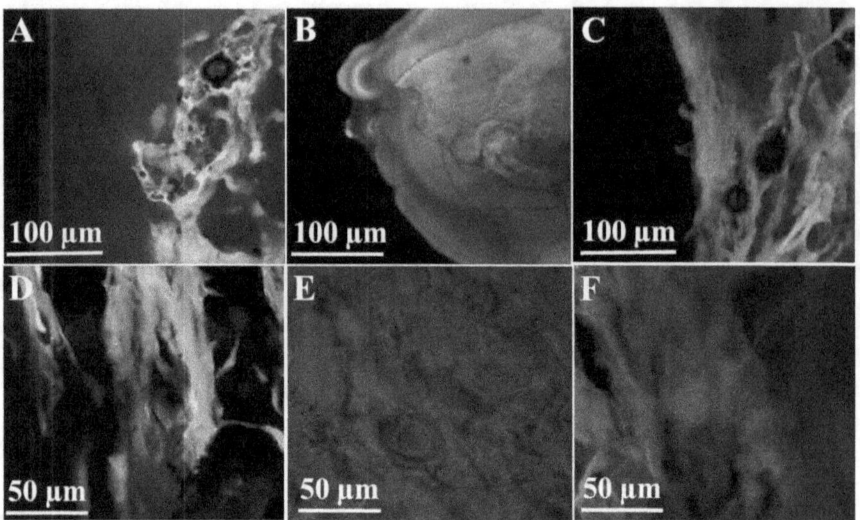

Figure. 6.14. Internalization of FITC from nanofibers mat in skin section by Franz type diffusion cells apparatus at different pH (A, D) pH 5.0 (B, E) pH 6.0 (C, F) pH 7.0. Maximum green fluorescence permeation in pH 5.0 denotes the pH responsiveness of nanofibers.

6.6.7. TEM analysis

After *ex vivo* skin permeation studies, the morphology of the PCPP-CA-PHM nanofibers observed by a TEM and presented in Figure. 6.15. A. In the 7 days post-treatment *ex vivo*, some of the nanofibers became enlarge, broken, and fused, but their basic structures were still retained. Nanofibers size gained, due to water engulfed in the pores, the degree of cross-linking was also affected, and size deformation. This also proved that a cross-linking reaction does not affect the morphology of the nanofibers. Dry nanofibers have less void space compare to the fiber after treatment of skin permeation studies. The nanofibers diameter and increases in void space proved that the water absorption capability of the nanofibers (Figure. 6.15. B). In Figure. 6.15. C, denotes the PTX-PCPP-CA nanomaterial core inside the PHM shell nanofibers, and controllable matrix hold an active targeting nanomaterial system for sustained delivery.

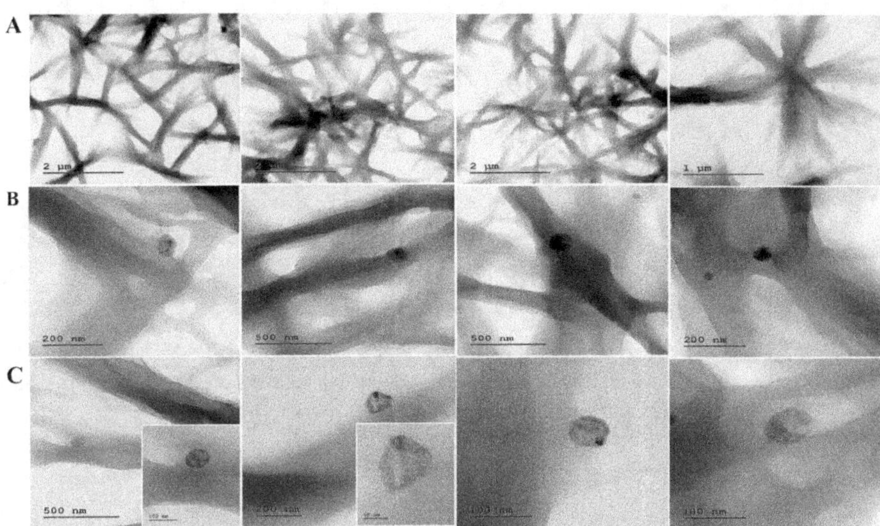

Figure. 6.15. PTX-PCPP-CA-PHM nanofibers degree of swelling were visualized by TEM image (A) Enlargement of nanofibers diameter, increases in void space between nanofibers (B) Images of core-shell nanofibers with nanomaterial (C) Presence of PTX-PCPP-CA nanomaterial inside core-shell nanofibers.

6.6.8. Cytotoxicity assays

MCF-7 cells cultured over the surface of PTX-PCPP-CA-PHM nanofibers were shown high cytotoxicity. On day 1, PTX-PCPP-CA-PHM nanofibers exhibited lower cell viability than the PTX; after 8 days, nanofibers were higher cytotoxic than that of the PTX in both cytotoxicity studies (Figure. 6.16. A, B). According to the *in vitro* drug release analysis, nanofibers initially burst release of PTX and controlled release for the long term, which helps in increases nanofibers cytotoxic after 8 days [187]. The free PTX displayed the lowest cell death at 8 days, which is ascribed to the free PTX diffuse quickly into tumor cells at an initial period and efficiency lost on day 8.

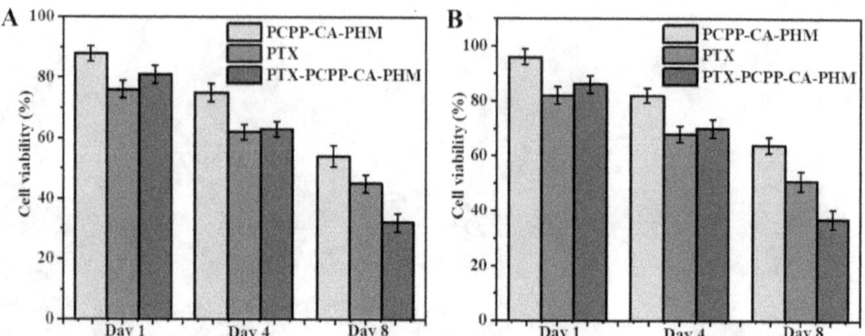

Figure. 6.16. (A) MCF-7 cell culture directly to PCPP-CA-PHM, PTX, PTX-PCPP-CA-PHM nanofibers viability was evaluated for days 1, 4, 8 (B) Nanofibers meshes were added, and fixed between two glass rings and viability was evaluated for days 1, 4, 8.

6.6.9. Flow cytometry

It was further confirmed by flow cytometry, PTX and PTX loaded nanofibers treated cells were used to measure viability, by gating the dead cells as per details mentioned in section 2.9.4 (Chapter 2). Control cells without treatment show a count of 12486 live cells without any peak shift (Figure. 6.17. A). PTX treated cells show the peak shift which indicates the formation of dead cells, and it around 60 % (Figure 6.17. B). Nanomaterial-loaded nanofibers have maximum peak shifts to R3 position, and 85 % of cells are dead (Figure 6.17. C). The sustained drug release ability of nanofibers is the reason to increases the apoptosis rate in cancer cells. Therefore, nanomaterial-loaded nanofibers had the best anticancer efficacy compare to PTX alone.

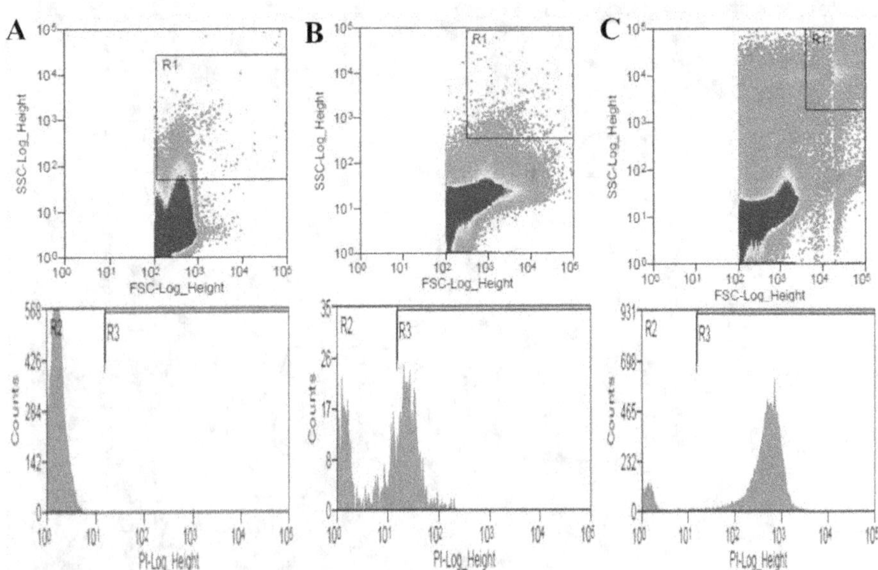

Figure. 6.17. Flow cytometry for identification of viable/dead (A) Quantitative analysis of cells without treatment and its histograms peak shift (B) Quantitative analysis of dead cells after PTX treatment and histograms peak shift (C) Quantitative analysis of dead cells after PTX-PCPP-CA-PHM nanofibers treatment and histograms peak shift.

6.6.10. SEM analysis

In SEM micrographs, MCF-7 cells were successfully seeded on PTX-PCPP-CA-PHM nanofibers, where the cells completely adhered to nanofibers. As shown in Figure. 6.18. A, B, cells hold into nanofibers, intact, and proper cell attachment. Higher cell spreading on nanofibers was observed after 24 h of cell culture. In Figure. 6.18. C, D, increase in time; nanofibers treated cells showed a change in the morphology, formation of apoptotic bodies, membrane blebbing, cell degeneration, and lysis [188].

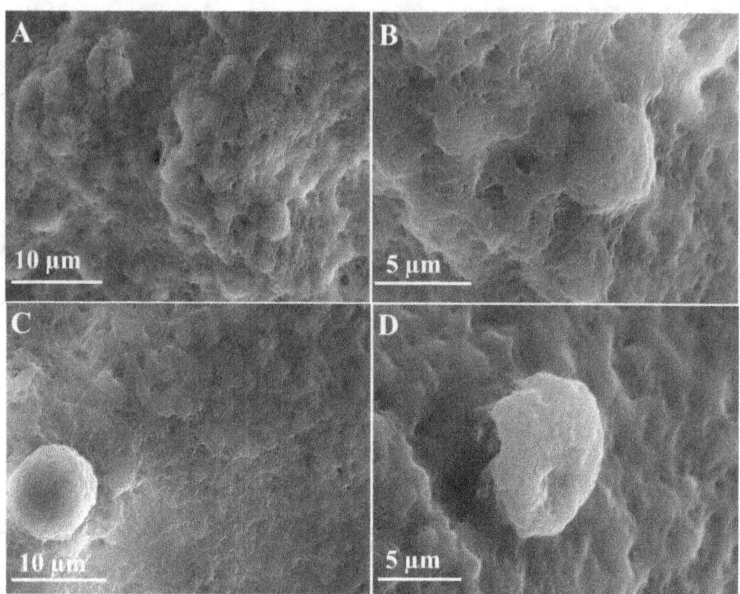

Figure. 6.18. SEM images of MCF-7 cancer cells seeded over PTX-PCPP-CA-PHM nanofibers for (A, B) 12 h (C, D) 24 h maximum treatment time clearly showing the apoptotic body formation and membrane blebbing in treated cells.

6.7. Conclusion

Active targeting CA conjugated nanomaterial loaded core-shell PHM nanofibers were prepared by coaxial electrospinning. PTX loaded into CA conjugated PCPP polymeric nanomaterial which was encapsulated into the nanofibers. The pH-responsive drug release of nanofibers exhibited intensely initial burst release followed by sustained release behavior for 180 h. In *ex vivo* skin permeation studies, nanofibers show high skin permeation and retention of the drug through Franz cell diffusion. The internalization ability varies based on pH; confocal microscopy analysis showed nanofibers mats could deliver the FITC into the skin, which was accumulated mainly in the epidermis. PTX eluted from nanofibers inhibits the growth of MCF-7 cancer cells through the generation of ROS and cell cycle arrest. This study indicated the fabrication of novel materials to ensure the stimuli-responsive controlled delivery of PTX and its potential application leads to safe cancer therapy.

7.1. Introduction

Tuberculosis is a bacterial disease caused by *Mycobacterium tuberculosis* and it is a high-risk pathogen due to the superficial adaptation to the intracellular milieu of alveolar macrophages in the respiratory tract. *M. tuberculosis* uses mannose, complement, and Fc receptors mediated enter into the macrophage and reside within a vacuole [189,190]. It enhances its intracellular survival by adapting to the metabolic machinery of the host cell. This localization of bacteria inside the cells protects them from lysosomal attack, humoral, and cellular immune response, and confines the efficacy of antibiotic treatments [191,192]. This reduces the efficiency of multiple drugs like RIF, IZN, ethambutol, and pyrazinamide. Liposomes are previously used to improve the bioavailability of drugs by increasing cellular delivery and enhances the pharmacokinetics at the alveolar macrophage site of infections [193]. However, the poor stability, passive diffusion through cell membranes for drug release are limit the efficacy. Liposomal surfaces decorated with active ligands were widely used for site-specific delivery of the drug into alveolar macrophages using receptor-mediated endocytosis. The arginine amino grafted PCPP was prepared and functionalized with liposomes to form arginine-g-PCPP polymer grafted liposomes (PGL).

7.2. Previous Work

Thus, various researchers proved the mannosylated liposomes exhibited enhanced cellular uptake associated with the mannose content on the liposomes. Similarly, alkylated bovine serum albumin specific affinity for macrophages scavenger receptors and which improves the progress of liposomes [194,195]. The immunomodulatory effect of the poly (organo) phosphazenes, was already reported and appears as a potent immunoadjuvant in nanovaccine delivery [196]. A study on animal models shows the PCPP significantly improves the immune response compared with empty bacterial and viral antigens. There were successful clinical trials in humans that represent the PCPP as an excellent safe, immunogenic property [197,198].

7.3. Novelty

Arginine serves as an immune-nutrient, modulator of the innate response, targeting ligand, production of nitric oxide, activation, and proliferation of macrophage [199]. This PGL improves the targeting of mycobacteria-infected macrophages via arginine-g-PCPP and increases penetration to the alveolar macrophage. The endosomal destabilization behavior releases the drug from liposomes, which increases the drug availability in the intracellular region of TB infected macrophage. Additionally, arginine-g-PCPP activates the inflammatory response, induces nitric oxide production, and enhances T-cell stimulation against the TB infected macrophage.

7.4. Objective

In this study, we prepared an arginine amino group grafted PCPP for targeting macrophages and trigger the inflammatory response. The study deals with structural orientation and interaction of the polymer and liposomes headgroup, which represents structural and physicochemical characterization. Dual drug loading and EE of arginine-g-PCPP-liposomes nanocarriers was studied. Evaluation of effective intracellular drug release based on the pH and immune-stimulating activity of liposome nanocarrier against *Mycobacterium tuberculosis*. Additionally, U937 cell lines were used to measure the proliferation rate of macrophages and identify the toxicity rate of arginine-g-PCPP-liposomes.

7.5. Experimental techniques
7.5.1. Synthesis of PCPP-*g*-arginine conjugates

PCPP-g-arginine synthesized by the amide bond formation of the carboxyl group of PCPP and amine group of arginine through the EDC/NHS coupling chemistry. The PCPP of 1 % wt mixed in MES buffer (pH 5.0). The carboxylic acid group of PCPP was activated by the addition of EDC, NHS, and the mixture was mixed for 2 h. Then, arginine (50 mg/mL) was added and kept for 24 h mixing. Hydroxylamine (50 mM) was added to the mixture to remove unbound molecules. Finally, PCPP-g-arginine washed several times using dehydrated ethanol.

7.5.2. Preparation of blank and dual drug-loaded liposome

Multifunctional liposomes were prepared using the lipid film hydration technique with slight modification from the earlier report [200]. Briefly, three different molar ratios (Table. 7.1) of lecithin and cholesterol were used to prepare the blank multifunctional liposomes. The mixture was dissolved in methanol: chloroform (1:2,v/v) mixture in a round bottom flask. The organic mixture slowly evaporated and thin lipids film formed by a rotary evaporator at 40 °C (BuchiVaccum controller V-850 Rotavapour-210). The lipid film was slowly hydrated with 5 mL PBS buffer (pH 7.4) overnight. RIF and IZN (5 mg/mL) are hydrophobic and hydrophilic antimycobacterial drugs used to prepare dual drug-loaded liposomes. The above-described similar procedure was followed for the preparation. The hydrophobic RIF was dissolved in an organic solvent with lipid content and hydrophilic IZN dissolved in PBS buffer. Further, the suspension was sonicated for 15 min and extruded at 60 °C through the polycarbonate filters. Later, the suspension was added to a dialysis bag and dialyzed against deionized water for 2 days to remove the unloaded drugs.

Table. 7.1. Composition of different liposomal formulations

Sample	Lecithin (Mole ratio)	Cholesterol (Mole ratio)
Liposome 1	7	4
Liposome 2	7	6
Liposome 3	7	8

7.5.3. Arginine-g-PCPP coated liposomes (PGL) preparation

Initially, 100 mg of arginine-g-PCPP polymer solution added dropwise to liposomal (100 mg) solution under magnetic stirring. The polymer, liposomes mixture was stirred for 1 h at 70 °C, and then, the unbound sample mixture was separated by centrifugation. Further, arginine-g-PCPP grafted RIF and IZN loaded liposomes were lyophilized by Sub Zero Lyophilizer, India.

7.6. Results and discussion

The arginine grafted PCPP polymer, and it coated on anti-TB drugs loaded liposomes for immune triggering and stimuli-responsive release of dual drug. Arginine-g-PCPP was successfully synthesized under a suitable condition and is characterized by ^{1}H-NMR spectroscopy (Figure.7.1).

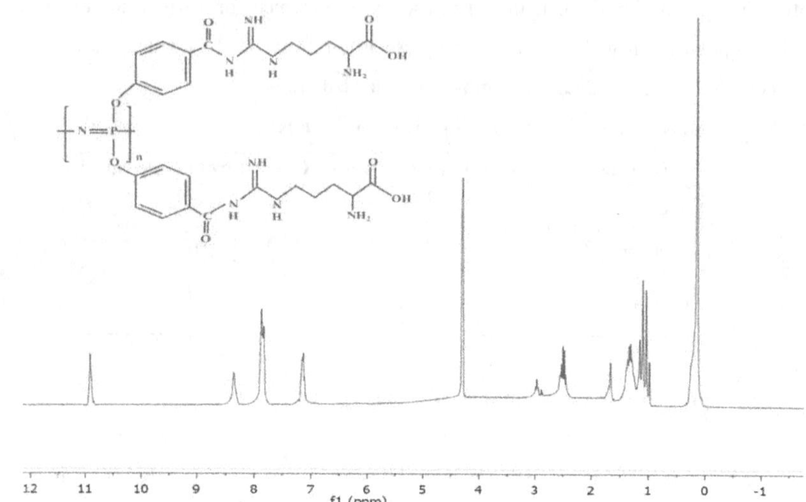

Figure. 7.1. ^{1}H NMR Spectra of arginine-g-PCPP in DMSO at 300 MHz.

7.6.1. Characterization of liposomes and PGL

(i) DLS and zeta potential of liposomes formulation

Liposomes formulated from lecithin/cholesterol and loaded with RIF and IZN by thin-film method. Initially, three different ratios of lecithin/cholesterol were studied concerning the surface charge, size, stability, and polydispersity index. The varying in cholesterol causes the change in liposomes characteristics significantly on an increasing size range. Liposome's concentration (7:4, 7:6, 7:8 - lecithin: cholesterol) exhibited the size range of 124, 323, 412 nm (Figure.7.2. A, B, C). In the molar ratio of 7:4, particle size is reduced, with narrow size distribution. The optimum concentration favors the uniform dispersion of lipid molecules to form nano-sized particles [201]. The polydispersity index value was 0.3 which shows the controlled size of liposomes with a narrow dispersity.

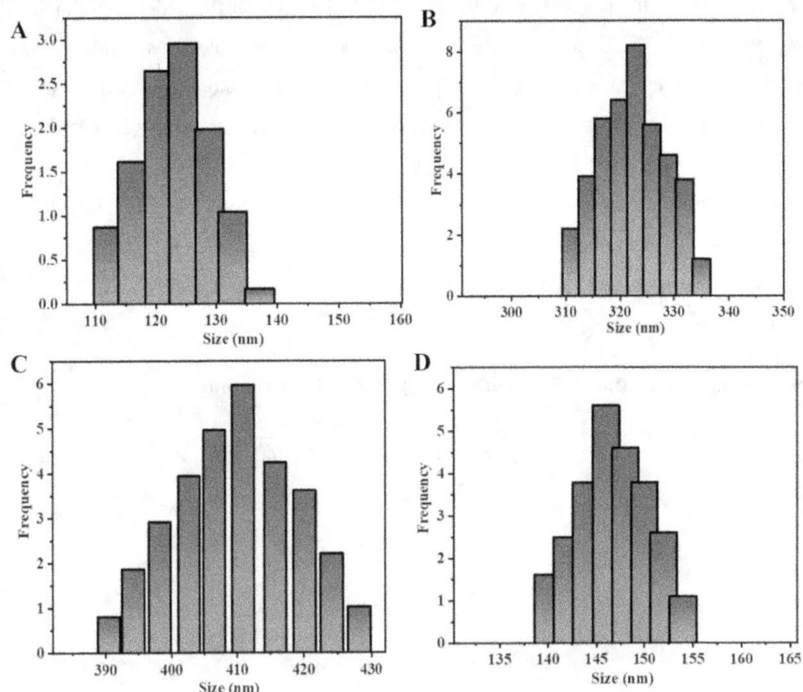

Figure. 7.2. DLS curve of (A) Liposomes (7:4) (B) Liposomes (7:6) (C) Liposomes (7:8) (D) Polymer grafted liposomes.

The optimum ratio of lecithin/cholesterol was chosen for further reaction with PCPP-g-arginine. Uncoated liposomes formed at a size of 124 nm, and after grafting with arginine-g-PCPP, the size increased to 146 nm (Figure.7.2. D). It shows an even distribution of polymer around the liposomes and the interaction formed by hydrogen bonding [202]. This association improves the stability of the PGL and dispersity. Most liposomes are unstable during the storage of drugs, which were protected by arginine-g-PCPP, and it also maintains the integrity of liposome structure. It formed optimum size, which is a vital characteristic of cell penetration and biodistribution. The average size range of 150-200 nm is an effective size for the drug delivery systems and helps the reticuloendothelial system escape [203,204]. The different molar concentrations of lecithin/cholesterol exhibit the zeta potential between −7.4 to −40.3. The negative zeta value was due to the phospholipid's head group. Usually, vesicles without cholesterol have weak electrostatic repulsive force. The incorporation of cholesterol elevates the negative zeta value, and its concentration finalizes the colloidal stability. As illustrated in Table. 7.2, a lower cholesterol ratio shows high stable liposomes compared to higher frequencies. Here, increasing the cholesterol concentration makes more hydrophobicity on the surface. The PGL shows -34.1 mV, which was adequate to maintain the dispersion's high stability due to the electrostatic interaction with polymer. The negative zeta values help in the prevention of PGL from aggregation [205]. Additionally, the negatively charged liposomes are effectively repulsed by the gastrointestinal tract lining which contains a negative value. It prevents the liposomes from insignificant absorption and induces the Na+ ion (physiological medium) binding to the phosphate oxygen group of liposomes [206].

Table. 7.2. Zeta potential value of different liposomal formulations.

Sample	Zeta value (mV)
Liposome 1	-40.3
Liposome 2	-25.1
Liposome 3	-7.4
PGL	-34.1

(ii) FTIR analysis

The synthesized polymer, liposome, polymer-coated liposome, and dual drug-loaded polymer-coated liposome were characterized using FTIR spectroscopy. The wavenumber 1425, 1686, 773, 847, and 613 cm^{-1} indicated the aromatic C=C stretching, –OH, C=O stretch, COO$^-$, C=C stretch of PCPP polymer. The adsorption bands at 2926 cm^{-1}, 1326 cm^{-1}, 1139 cm^{-1} are assigned to CH_3 asymmetric stretching, CH bending, CN stretching of arginine, respectively, and the arginine grafted on PCPP polymer (Figure. 7.3. A). The liposomes showed the absorption bands at 3186 cm^{-1}, 1810 cm^{-1},1680 cm^{-1}, 1510 cm^{-1},1140-1210 cm^{-1}and 1465 cm^{-1} represents the aliphatic ester stretching, CO, PO_2 groups stretching, CH_2 stretching. The absorption band 3350 cm^{-1} and the range between wavenumbers 3750 cm^{-1} and 3550 cm^{-1} represents the stretching vibration of hydroxyl and amine groups. The spectra between the 1650 cm^{-1} and 1700 cm^{-1} are related to the vibration of C-C, indicating the strong interaction between lecithin and cholesterol (Figure. 7.3. B). PGL is expanding the peak at range 3500 cm^{-1}, which denotes the strong interaction. The spectra in 1700 cm^{-1} are related to aromatic C-C of PCPP, and 1210 cm^{-1} denotes the PO_2 of liposomes. PGL exhibited bands for different reactive functional groups, like $-NH_2$, -OH, COOH, and CO groups, due to strong interactions between the functional groups of the liposomes and arginine-g-PCPP (Figure. 7.3. C). It indicates the formation of weak hydrogen bonds, Vander Waals forces, dipole-dipole interaction, etc. This interaction of polymer carboxyl group and a head group of lipid layers confirms the strong association. It also suggests the arginine-g-PCPP role interaction with other lipids headgroups on cells to improve the cell internalization [207].

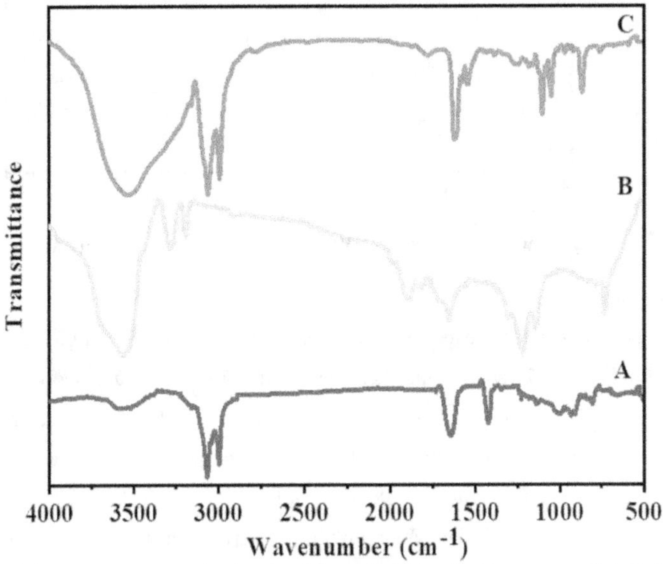

Figure. 7.3. FTIR spectroscopy of (A) Arginin-g-PCPP (B) Liposomes (C) Polymer grafted liposomes.

(iii) XRD pattern

Investigate the crystalline or amorphous nature of liposomes and PGL was examined by X-ray diffraction. In Figure.7.4. B. denotes the 2Θ values of liposomes at the range $21°$ represent the short spacings of fatty acid chains and possess an amorphous nature. Arginine-g-PCPP exhibited the sharp, intense peak that denotes the crystalline form. Arginine-g-PCPP coated liposomes clearly show no crystalline peaks (Figure.7.4. A). The PGL were broader, with lower intensities compared to the liposomes. The broad peak was due to the crystal effects of polymer conjugates (Figure.7.4. C).

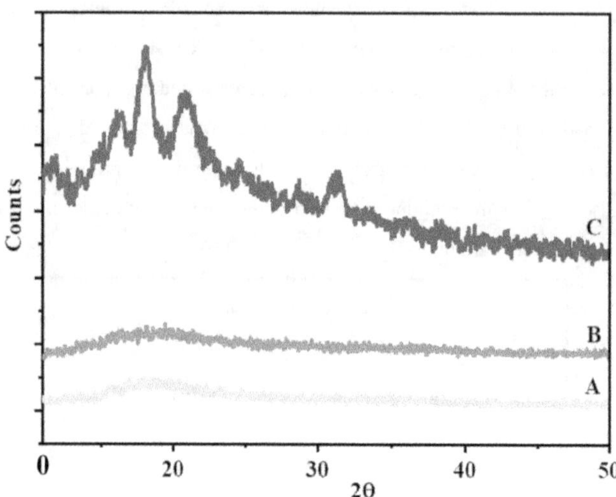

Figure. 7.4. XRD pattern of (A) Arginin-g-PCPP (B) Liposomes (C) Polymer grafted liposomes.

(iv) TGA studies

The TGA graph represents the thermal stability of liposomes, arginine-g-PCPP, and PGL, represented in Figure.7.5. The thermal behavior varied from PGL and liposomes were due to the arginine-g-PCPP. The liposome rate of weight loss was high in lesser temperature, showing a 40 % weight loss at 150 °C. Arginine-g-PCPP and PGL show similar weight loss and the major mass loss difference of liposome was due to the water content level. PGL, initial weight loss was occurred from 180 to 190 °C, because of H_2O elimination from the material. Compare to the initial weight loss; there was a maximum loss that occurred at the second stage with major elimination of CO_2, CO, NO_2, and other oxide molecules. Polymer interacts with the phospholipid head group of liposomes by the formation of the hydrogen bond. The rapid breakdown of interaction between polymer and lipid causes the maximum loss in the second temperature. Polymer coating triggers a shift in degradation temperature and 50 % weight loss was maintained above 340 °C. The higher thermal stability of liposomes was due to the polymer matrix coating.

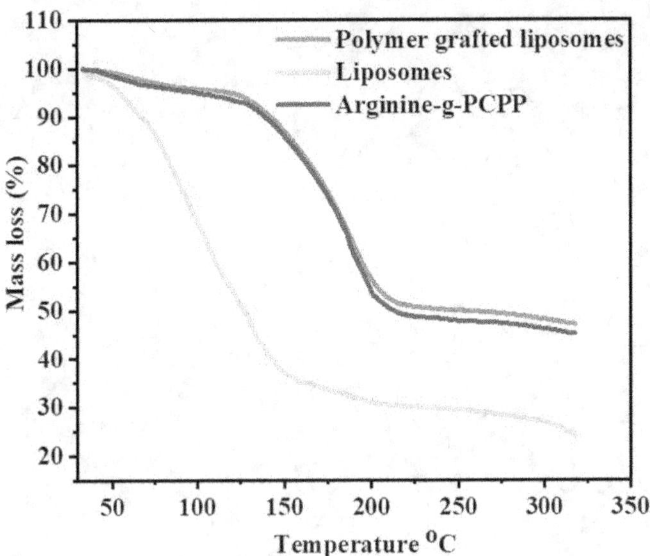

Figure. 7.5. TGA curve analysis of Arginin-g-liposomes, Liposomes, Polymer grafted liposomes.

(v) SEM analysis

Liposomes were observed under a scanning electron microscope to characterize the shape and morphology. In Figure.7.6. A-D, confirms the formation of nanosized and spherical liposomes, which correlate with the DLS results. It forms well dispersed, integrated, homogenous, and spherical liposomes. Liposomes form by a lipid layer, which represents the large multilamellar vesicles.

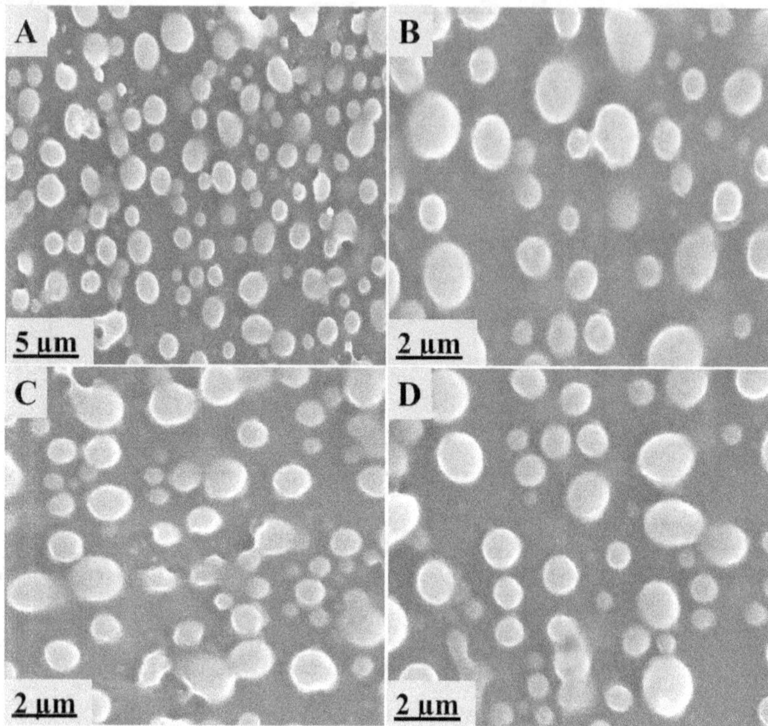

Figure. 7.6. SEM image representing the structure and morphology of liposomes (A) 5 µm (B) 2 µm (C) 2 µm (D) 2 µm.

Lecithin/cholesterol at the ratio 7:4 has spherical and stable morphology, whereas the liposomes formed at the 7:6, 7:8 ratio are irregular (Figure. 7.7. A, B). In both concentrations, it forms an irregular shape, agglomerate, and no clear structure.

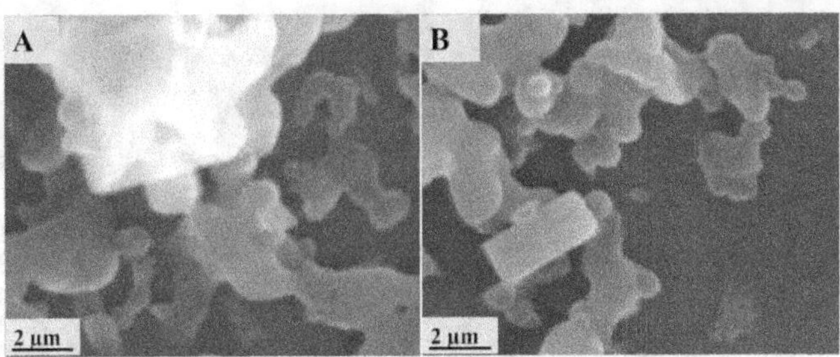

Figure. 7.7. SEM image representing the structure and morphology of liposomes prepared by lecithin/cholesterol (A) Molar ratio (7:6) (B) Molar ratio (7:8).

Liposomes with a smaller size range have more efficiency in drug delivery compared with larger-sized liposomes. But they were not suitable for long circulation, and it has limited use as it was trapped by the mononuclear phagocyting system [208]. To avoid phagocytic trapping and to target liposomes to the TB-infected macrophage site, it was grafted by an arginine-g-PCPP polymer. PGL also shows spherical-shaped large multilamellar liposomes represents in Figure. 7.8. A,B. The average liposomal size after coating was found to be 140 nm. The size is similar to the size obtained by the dynamic laser light scattering technique. The arginine-g-PCPP polymers have biodegradable, biocompatibility, mucoadhesive, and alveolar macrophage targeting properties. And, the polymer-bound to liposomes by hydrogen bonding because of its abundant source of hydroxyl/oxo groups. The liposomes hydroxyl group function as a hydrogen-bonding donor and the arginine-g-PCPP oxo groups are act as hydrogen bonding acceptors. SEM image confirms the enhanced interaction, and it shows tight adsorption of polymer on the surface of the liposome. Wenjia *et al.,* demonstrated the polymer and lipid interaction formed by intermolecular hydrogen bonds within the hydrophilic head group contains hydroxyl groups [209]. Additionally, there is the formation of hydrophobic interaction with arginine-g-PCPP segments in the presence of cholesterol in liposomes.

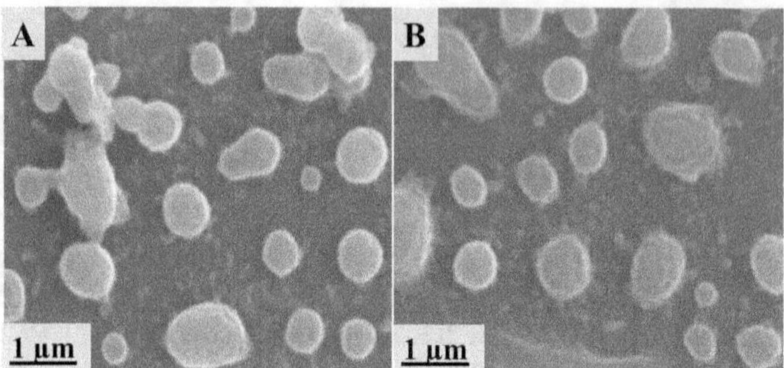

Figure. 7.8. (A,B) SEM image representing the shape of polymer grafted liposomes.

(vi) TEM analysis

TEM was used to visualize the grafting ability of arginine-g-PCPP on the surface of the liposome (Figure. 7.9. A-F). TEM images indicate the different forms of the structure formed due to the arginine-g-PCPP coating on the liposome surface. The polymer is effectively embedded on the liposome surface, and the arrangement of arginine-g-PCPP slightly alters the liposome structure. But mostly, it was spherical-like morphology and a highly stable structure. TEM results indicated the lesser size range of liposomes compared to DLS data. It may due to the dry liposomes were used for TEM analysis. In DLS analysis, the size obtained was the hydrodynamic diameter of liposomes in the buffer. More amount of water and some ions molecules associated with the buffer increase the size of liposomes. Arginine-g-PCPP dictates the binding interaction with lipids to represents the bi-block structure on the outer surface of liposomes. The rich functional groups allow strong interaction with the pristine structure.

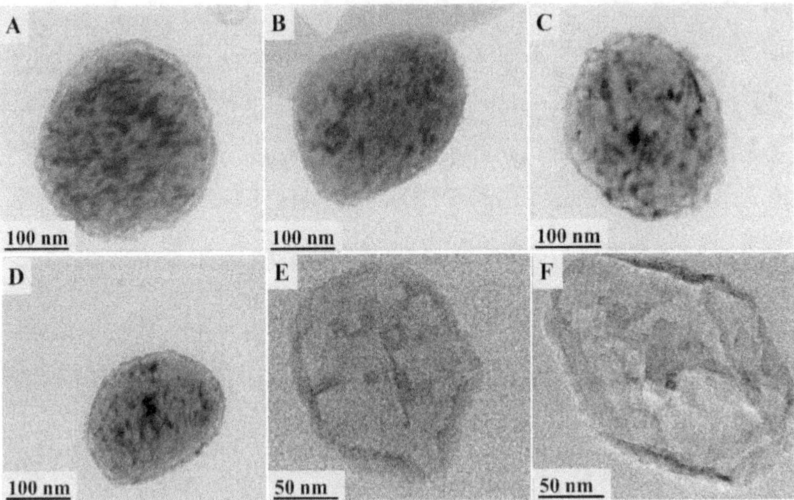

Figure. 7.9. TEM image of polymer grafted liposomes in different magnification (A-D) 100 nm (E-F) 50 nm.

7.6.2. Encapsulation and loading efficiency

Encapsulation and loading efficiency of PGL ability depended on several factors. Significantly, high drug encapsulation and loading of the hydrophobic and hydrophilic drugs were achieved by the multifunctional liposome. Maximum absorbance intensity was obtained at a drug concentration of 4 and 12 %. Both RIF-IZN loaded PGL shows decreases in absorption intensity at 8 % mol concentration, which indicates the effective loading of drugs. The percentage of drug-loaded on liposomes was 84.21 % (RIF) and 81.34 % (IZN) in 8 % concentration (Table.7.3). The EE of both drugs was also successful in 8 % concentration. The EE was maximum of 91.63 %, 88.25 % at 8 % the concentration of RIF and IZN. It observes a decrease in encapsulation in 4 and 12 % concentrations. An excessive amount of drug concentration competes for a position in liposomes and decreases in encapsulation. Higher encapsulation of drug is due to the interaction between the hydrophobic core of lipid composition and hydrophilic drug encapsulated into the aqueous compartment of liposomes [210].

Table. 7.3. RIF and IZN drug loading and encapsulation efficiency of Polymer grafted liposomes.

Drug Mol %	Loading efficiency		Encapsulation efficiency	
	RIF	IZN	RIF	IZN
4	75.252	72.456	66.360	63.216
8	91.634	88.253	84.210	81.34
12	83.235	80.356	72.141	70.524

7.6.3. *In vitro* drug release

The pH-responsive drug release was observed in RIF/IZN drugs loaded liposomes and PGL nanocarrier in two different pH conditions at 37 °C. The different pH condition-based study relates to the pH gradient in endosomal-lysosomal trafficking and to prove the PGL preferential drug release at the cytoplasm region. Initially, both drugs exhibited the burst release at 12 h for both pHs such as 5.0 and 7.0 (Figure.7.10). Compare to the liposomes, PGL exhibited the controlled release of drugs over 140 h with a maximum release of 88 %. Additionally, PGL exponentially releases both drugs for 100 h and which is attributed to the slow detachment of drugs from nanocarrier. Both drugs presented slight variation in release profiles for 140 h. PGL exhibited pH-based drug release of 88 %, and 84 %, RIF (Figure.7.10.A), and IZN (Figure.7.10.B) at pH 5.0. In pH 7.0, drug release of 63.0 % and 60.0 % RIF (Figure.7.10.A) and IZN (Figure.7.10.B), which comparatively less. *In vitro* drug release confirms the dual drug-loaded PGL exhibit a higher release profile at acidic pH, which represents the more release during endosomal acidification inside the TB infected macrophage cell. The difference in drug release rate between the acidic and neutral pH is mainly due to the arginine-g-PCPP. This controlled pH-responsive drug release profile delivers the maximum drug at the intracellular infection site with longer therapeutic potential and lower frequency of drug administration [211].

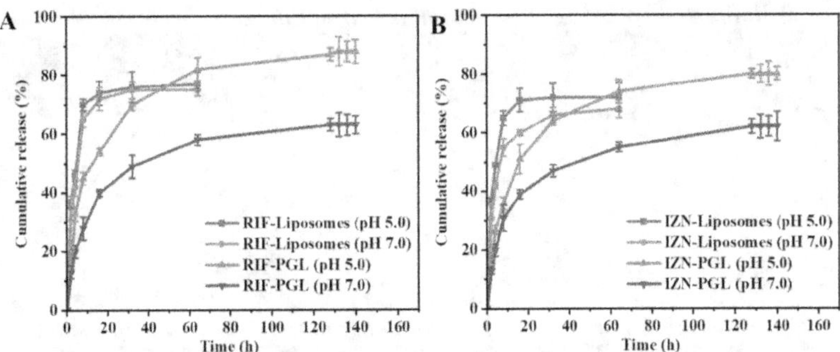

Figure. 7.10. (A) *In vitro* **cumulative drug release of RIF from liposomes and polymer grafted liposomes (B)** *In vitro* **cumulative drug release of IZN from liposomes and polymer grafted liposomes under different pH.**

7.6.4. Antimicrobial activity

The antimicrobial effect of RIF and IZN loaded PGL was investigated against both Gram-positive (*Bacillus subtilis, Staphylococcus aureus, and Clostridium sp.*) and Gram-negative (*E.coli, Klebsiella pneumoniae, Salomonela typhi*) bacteria using the disc diffusion method (Figure. 7.11). The zone of inhibition, MIC, and MBC of the free drugs and drug-loaded PGL has observed an increase in fold area was calculated. The antibacterial effect varied on the different species; RIF antibacterial activity was less on all bacterial pathogens due to the poor solubility. IZN exhibited an effective antibacterial effect compare to the RIF due to its solubility nature. RIF synergistic effect increased after loaded with the liposomes; it improves the interaction time with bacterial membrane and an increase in effective tendency against all pathogens.

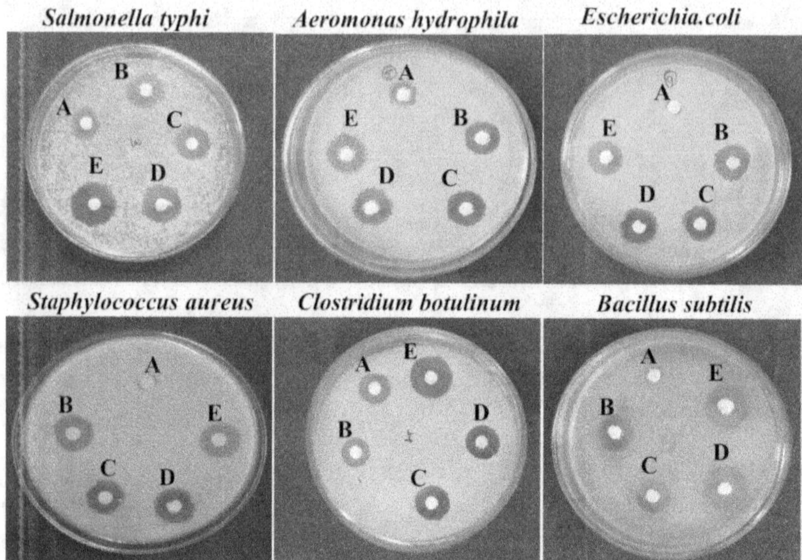

Figure. 7.11. Antimicrobial effect of different strains of gram-positive and gram-negative bacteria against (A) RIF (B) IZN (C) RIF loaded PGL (D) IZN loaded PGL (E) RIF-IZN loaded PGL.

The combination effect of drugs is the reason for the better MIC, MBC, and zone of inhibition values. In Table 7.4, which represents the excellence of PGL in the drug delivery system. The RIF and IZN loaded PGL shows an excellent action compared with free RIF and IZN, RIF-PGL, and IZN-PGL. The MIC and MBC values of dual drug-loaded PGL were significantly lower than the other formulation in all pathogens. Accelerated protonation causes the breakdown of the ester bond, which destabilizes the PGL and triggers drug release [212].

Table. 7.4. Antimicrobial effect of different strains of gram-positive and gram-negative bacteria against RIF, IZN, RIF loaded PGL, IZN loaded PGL, RIF-IZN loaded PGL, and its MIC, MBC and Zone of inhibition.

	S. typhi			*K. pneumoniae*			*E.coli*			*Clostridium sp.*			*B. Subtilis*			*S.aureus*		
	MIC	MBC	ZI	MIC	MBC	ZI	MIC	MBC	ZI	MIC	MBC	ZI	MIC	MBC	ZI	MIC	MBC	ZI
A	18.35	26	13	23.78	34	11	27.26	38	8	20.13	32	14	27.86	39	6	24.12	33	7
B	12.52	20	14	15.21	23	11	19.25	27	15	12.45	20	15	21.26	18	12	16.03	23	15
C	13.26	22	16	17.26	25	15	20.22	29	17	14.01	22	16	22.36	30	14	18.45	28	15
D	8.43	15	17	11.21	21	17	15.52	29	18	9.35	20	18	18.21	29	15	12.03	23	16
E	1.26	3	20	3.01	6	18	4.21	7	19	1.46	2	19	5.82	8	16	3.23	6	18

A= RIF, B= IZN, C= RIF-polymer-liposomes, D= IZN-polymer-liposomes, E= RIF-IZN-polymer-liposomes

7.6.5. Antimycobacterial activity

To assess the antimycobacterial effect of RIF, IZN, RIF loaded PGL, IZN loaded PGL, and RIF&IZN loaded PGL were performed growth inhibitory assays (Figure. 7.12). Bacterial growth is determined by the reduction of absorbance value OD at 600 nm. The result shows, free RIF, IZN do not effectively inhibit the *M.tuberculosis* compare to drug-loaded polymer-coated liposomes. Single drug-loaded PGL also shows effective inhibition compared with RIF and IZN. The RIF and IZN loaded polymer- liposomes inhibited *M.tuberculosis* growth with increasing dose and since the activity is dose-dependent. Consequently, the effective therapeutic action of drug-loaded PGL was due to the cleavage of hydrogen bond interaction of arginine-g-PCPP and liposomes. The bacterial growth inhibition effect of dual drug-loaded PGL indicates the combination of different TB drugs can strongly inhibit *M. tuberculosis*. Importantly PGL helps in good distribution around the bacteria surface and helps in ease binding with the help of arginine. Bacterial cell walls are mostly negative, positive amino groups of PGL form polyelectrolyte complex with the bacterial surface. Higher PGL internalization of the macrophage cell membrane and endosomal drug release ability of PGL increases the RF and IZN concentration.

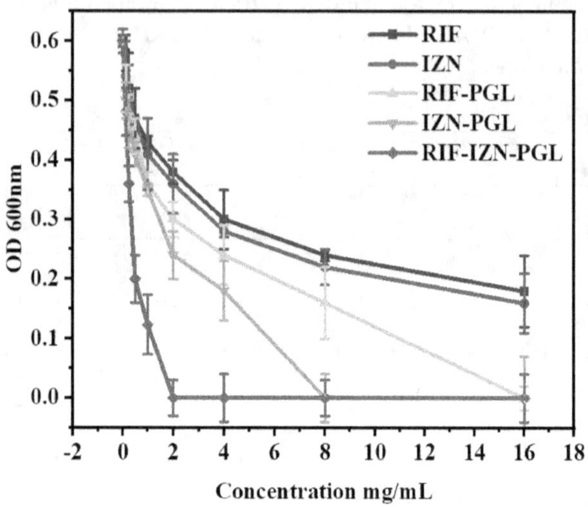

Figure. 7.12. *Mycobacterium* **growth inhibitory effect of RIF&IZN loaded PGL.**

7.6.6. *In vitro* biocompatibility studies

The cytotoxicity of RIF, IZN, RIF, and IZN loaded PGL was studied against U937 cells by MTT assay (Figure. 7.13). The good cytotoxic profile of RIF and IZN shows the side effect increased with rising concentration. The 50 µg/mL concentration of RIF and IZN exhibited 40 % of cell survival and denoted the toxic effect of drugs. So, the combination of multiple TB drugs for treatment is dangerous by causing damage to normal cells and possibly developing drug resistance to TB. In this case, drug-loaded liposomes were not much toxic to the cells; particularly, dual drug-loaded liposomes did not affect cells at a higher concentration of 50 µg/mL. There was 40 % of cell death at the concentration at 50 µg/mL drug-loaded liposomes. This toxic effect also due to the toxic profile of drug leakage on the PGL moiety. The drug-loaded liposomes showed a positive impact on cells, representing their capability to control the side effects of RIF and IZN. Similarly, polymer grafting endows the liposomes to control the drug release in the non-specific region.

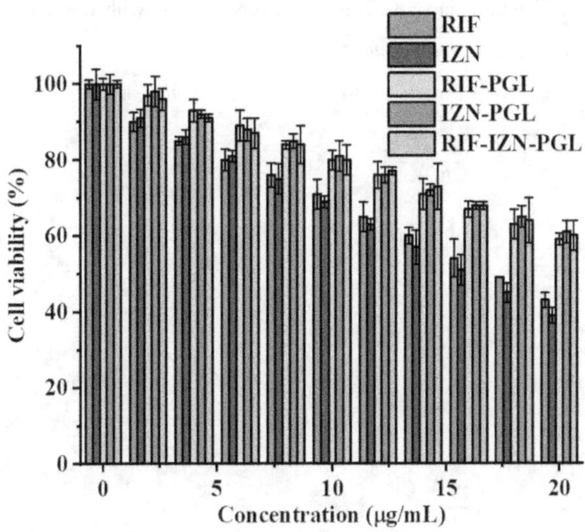

Figure. 7.13. Biocompatible effect of RIF and IZN loaded PGL.

7.6.7. Macrophage cell proliferation

The U937 macrophage cells were cultured with liposomes and arginine-g-PCPP coated liposomes to evaluate the arginine efficiency of cell proliferation. In Figure.7.14, the phase-contrast microscope image represented the samples with four different concentrations (20, 30, 40, and 50 µg/mL) and visualized after 24 h. Plain liposomes added to the cells shows less proliferation which was similar in all four concentrations. But, arginine-g-PCPP coated liposomes, exhibited an increased number of cells with raising the concentration. In 20 µg/mL concertation, cells were more, and in comparison, to liposomes alone, treatment with similar concentration shows a high proliferation rate. Followed by raising the concentration shows increasing in cell number. This difference was due to the arginine amino acid on the polymer. Arginine is essential for macrophage cell metabolism, and it forms proline to induces cell proliferation. Additionally, it produces polyamines for tissue repair and nitric oxide, which induces inflammation. Several studies reported the arginine helps in the proliferation of macrophage and nitric oxide production [213]. So, the presence of arginine in PGL effectively induces antimycobacterial activity by activating the innate immune system, which has a vital role in the control of inflammatory reactions. Overall, PGL was a good nanocarrier with excellent antimycobacterial efficiency, and this is the first report deals with immune-modulating nanocarrier able to deliver dual drugs for *M. tuberculosis*.

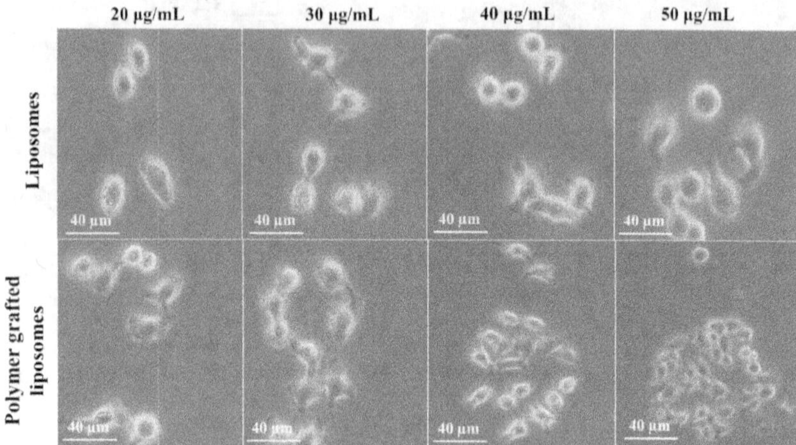

Figure. 7.14. U937 macrophage cells proliferation effect of liposomes and polymer grafted liposomes in different concentrations.

7.7. Conclusion

This study confirms the utility of arginine-g-PCPP to functionalize the liposomes for immunomodulating and intracellular delivery. Different liposomes formulation is utilized to identify the optimum sized liposomes. The characterization results concluded the effective interaction of arginine-g-PCCP and lipid headgroup. Biointerface interaction between lipid and polymer was due to the presence of various functional groups that form a hydrogen bond, Vander Waals force, etc. The arginine-g-PCPP significantly changes the drug release behavior of the liposomes. PGL shows endosomal pH-induced enhanced drug release inside the cell. Arginine activates the macrophage by increasing cell proliferation and more nitric oxide production. Therefore, a liposomal delivery system with immune-stimulating ability is a good approach, deliver both drugs to the cytosol where *M. tuberculosis* survives.

8.1. Introduction

The development of an alternative therapeutic strategy facilitates overcoming the drug-resistant TB emergence. Currently, phototherapeutic techniques like PDT and PTT fascinate intensive interest, since it is non-invasive, selective, effective and inexpensive technique. Majorly, it does not cause any resistance to drugs, and also it will not be influenced by the drug resistance status of the *Mycobacterium tuberculosis* [214]. PDT is a promising approach to kill the TB bacteria, under light irradiation which causes excitation of the photosensitizer. It will undergo singlet-to-triplet intersystem crossing and activation of oxygen to yield ROS. It mainly affects the bacterial cell membrane lipids, proteins, and nucleic acid leading to death [215,216]. So, an alternative technique, PTT is also an efficient approach like PDT. The photothermal agents triggered by light cause the thermal effect through non-radiative transition and increases in local temperature, which could damage the cell membrane structure, deactivate DNA, protein, and organelles [217].

ICG is an amphiphilic, tricarbocyanine organic dye, near-infrared (NIR) agent, with a strong absorption band at 780 nm and active emission band at 800-820 nm. The major disadvantages of ICG are aggregation, poor stability, and rapid clearance from the body. A combination of ICG with photoactive nanoagent can able to overcome this issue [218,219]. Some examples of photoactive nanoagent, are GO, rGO, MoS_2, WS_2, $MoSe_2$, WSe_2, etc offer numerous advantages. It has strong absorption in the NIR region, good spatial resolution, specific targeting, the tissue penetration depth of NIR lasers, more photothermal conversion during the PTT process. Generally, graphene and its derivatives rGO have a strong ability to absorb NIR at 700-1100 nm to release heat. rGO gains advantages like exhibits ultrahigh thermal conductivity, fewer side effects, larger surface area, chemical stability, and drug carrier [220-222]. MoS_2 layer-by-layer nanostructures composed of highly stable covalent bonds, ultra-high surface area, drug carrier, biodegradable, and good absorption in the NIR region.

8.2. Previous work

Several natural chromophores or photosensitizers like porphyrin, phenothiazine, cyanine, oligomers, and conjugated polymers were used in PTT and PDT [223]. In this, US-FDA approved indocyanine green (ICG) widely used in biomedical applications. Majorly, ICG was widely used for combined PDT–PTT, when ICG upon irradiation at 808 nm, it generates heat for PTT and singlet oxygen for PDT. Numerous studies reported the combination of graphene, its derivatives rGO with ICG into a single system shows effective phototherapies against cancer and bacteria [224,225].

8.3. Novelty

MoS_2 was extensively utilized photocatalytic and rarely used as an antibacterial agent. But it able to produce heat upon NIR irradiation. The bandgap of MoS_2 monolayer is 1.8 eV, indicating that the MoS_2 ability to higher photothermal conversion. It has low cytotoxicity, good charge-transfer property between graphene and MoS_2. So, a combination of MoS_2 and graphene can significantly increase the photothermal activity. In addition, it combines with PCPP to protect the ICG which generates the heat and singlet oxygen to eliminate the bacteria [226,227].

8.4. Objective

The fabrication of rGO-MoS_2 with PCPP polymer to carry ICG and antimycobacterial drugs. This photoactive nanoagent performs PTT, PDT, and chemotherapy against *M.tuberculosis*. Initially, PCPP was functionalized with antimycobacterial drug-isoniazid (IZN). Next, the synthesis of layered rGO-MoS_2 by L-cysteine assisted solution-phase method and layered 2D nanostructure facilitates the loading of ICG. Further, PCPP-IZN assembled on rGO-MoS_2-ICG to improve the immobilization and stability of ICG to form an rGO-MoS_2-PCPP-IZN-ICG nanocomposites. This all-in-one therapeutic nanocomposite triggered by NIR-light irradiation elicits PTT and PDT. The nanocomposites, drug, and ICG internalize into *Mycobacterial* cells, as well as hyperthermia, singlet oxygen generation by the short NIR irradiation. Our studies indicate the combined effect of rGO-MoS_2 and ICG has the potential to deliver effective treatment against *M.tuberculosis* via chemo/PTT/PDT technique.

8.5. Experimental techniques
8.5.1. Synthesis of graphene oxide

Graphene oxide was synthesized from natural graphite powder using the modified Hummers method [228]. Graphite powder (5 g) was slowly added to the concentrated H_2SO_4 under an ice bath. Then, $KMnO_4$ (3 g) was slowly added, stirred for 2 h, and diluted deionized water. After that, H_2O_2 (10 mL) of added to the solution until the color changed to brilliant yellow. The graphene oxide redispersed in water, further filtered, and exfoliated by ultrasonication to form graphene oxide sheets.

8.5.2. Preparation of rGO-MoS₂

The rGO-MoS_2 was prepared by a modified cysteine-assisted solution-phase method [229]. The prepared GO added into 50 mL water, along with 0.5 g $Na_2MoO_4 \cdot 2H_2O$, the mixture was stirred and ultrasonicated for 20 min. The solution pH was adjusted to 6.5 and it was added to 1 g of L-cysteine along with 100 mL water. Then, it transferred to the Teflon-lined stainless-steel autoclave and heated at 180 °C for 36 h. After cooling, black precipitates were separated by centrifugation washed with water and ethanol. The nanomaterial dried in a vacuum oven at 80 °C for 24 h and it annealed in the furnace at 800 °C for 2 h.

8.5.3. Functionalization of PCPP with IZN

Antimycobacterial drug isoniazid was functionalized with polymer. PCPP (1 %) mixed with MES buffer, and EDC, NHS was added. Isoniazid (20 mg /mL) added to the mixture and incubated for 24 h. The unreacted molecules were removed by hydroxylamine (50 mM) and washed with dehydrated ethanol.

8.5.4. Fabrication of rGO-MoS₂-PCPP-IZN

The rGO-MoS₂-PCPP-IZN nanocomposites were fabricated by gel casting technique [230]. For the preparation, PCPP-IZN of different concentration (1 to 4 %) was mixed with 50 mg of rGO-MoS₂ and mechanical stirrer at 2000 rpm for 2 h. Nanomaterial suspension was cast in mold and kept for drying at 40 °C. Then nanomaterial was peeled out and kept for drying in vacuum oven.

8.5.5. ICG loading into rGO-MoS₂-PCPP-IZN

Aqueous ICG solution (5mg/mL) was added to the rGO-MoS₂-PCPP-IZN nanomaterial suspension of 5 mL. The resulting mixture was a mechanical stirrer at 5 °C,12 h in dark. The ICG loading ability was evaluated by, separation of ICG from nanomaterial by repeated filtering through a 30 kDa filter and followed centrifuge at 10,000 rpm for 10 min. It was washed with water and unloaded ICG was measured by UV-vis spectroscopy at 780 nm. ICG loading ratios were determined by,

$$\text{ICG loading efficiency } \% = \frac{(\text{Weight of the loaded ICG})}{(Weight\ of\ the\ initial\ ICG)} \times 100$$

8.5.6. *In vitro* photothermal and photodynamic effect

The *in vitro* PTT under NIR laser irradiation was performed at 808 nm for the PCPP-IZN, GO, rGO-MoS₂, rGO-MoS₂-PCPP-IZN, and rGO-MoS₂-PCPP-IZN-ICG. The temperature change was measured by varying the time. To determine the production of singlet oxygen rGO-MoS2-PCPP-IZN, rGO-MoS₂-PCPP-IZN-ICG were mixed with N,N-dimethyl-4- nitrosoaniline (RNO), and imidazole in 10 mM PBS (pH 7.4). The resulting mixture was illuminated NIR light at 808 nm with a power of 1.1 W cm-2 with different time intervals. The RNO reduction was absorbed in 440 nm which denotes the single oxygen production.

8.5.7. Antibacterial effect of rGO-MoS₂-PCPP-IZN-ICG

Briefly, bacterial suspension of *E.coli* (Gram-negative), *B.subtilis* (Gram-positive)-100 µL with a concentration of 1.5×10^6 CFU/mL was added to the sterile 96-well plate. 100 µL PCPP-IZN, GO, rGO-MoS₂, rGO-MoS₂-PCPP-IZN, and rGO-MoS₂-PCPP-IZN-ICG were added to each well. The 96-well plate was maintained in a dark at 37 °C and then it was subjected to the laser illumination for 5 min. The number of CFU/mL was measured and the bacterial viability was also evaluated by plating 10 µL on a nutrient agar plate and incubated at 37 °C for 24 h. Bacterial morphology change was evaluated by a phase-contrast microscope in a different time interval.

8.5.8. Morphological characterization of *Mycobacterium* after irradiation

Initially, *Mycobacteria tuberculosis* suspension of 1 mL mixed with 5mg/mL of rGO-MoS₂-PCPP-IZN-ICG and expose to NIR-irradiation for 5 min. Then treated bacterial suspension was centrifuged at 13000 rpm for 10 min. Then, it was fixed with 1 % glutaraldehyde, 2 % formaldehyde, 4 % osmium oxide and then centrifuged. The obtained pellet was washed with deionized water and then resuspended in 1 mL of deionized water. Finally, morphological changes were observed in SEM (VEGA3SB, TESCAN).

8.6. Results and discussion

All-in-one nanomaterial-rGO-MoS₂-PCPP-IZN-ICG were prepared by solution-phase method and gel-casting technique. In this, nanomaterial exposure to the NIR-irradiation triggers the singlet oxygen production and hyperthermia which able to kill the TB containing granuloma. Additionally, nanomaterial able to penetrate the granuloma cells and release the IZN drug which blocks the InhA. Importantly, this process stops the mycolic membrane bio-synthesis.

8.6.1. Characterization of rGO-MoS₂-PCPP-IZN-ICG
(i) FTIR analysis

In Figure. 8.1, the FTIR spectrum of GO, rGO-MoS$_2$, PCPP-IZN, and rGO-MoS$_2$-PCPP-IZN were compared between the 4000 to 400 cm^{-1}. GO exhibited the bands in 3480 cm^{-1} represents the graphene with oxygen groups (Figure. 8.1.A). The presence of the oxygen-containing groups indicates the graphite has been oxidized. It also helps in hydrogen bond formation between graphite and hydroxyl groups. The band 1580 cm^{-1} attributed the C=C skeletal vibration, which confirms the transition of graphite to graphene oxide. In rGO-MoS$_2$, the bands at 1150, 1060 cm^{-1} correspond to the S-O stretching, and another sulfate species peak at 530 cm^{-1} resembles the disulfides stretch (S-S) (Figure. 8.1.B). The 3480 cm^{-1} band intensity was reduced compared to GO which indicates the changes in -OH levels. In PCPP-INZ, bands appear at 3010, 1770, 670 cm^{-1} represents the aromatic C-H stretching, aromatic C=C bending, C-H vibration, and 3520 cm^{-1} confirms the N-H stretching between PCCP, IZN (Figure. 8.1.C). In rGO-MoS$_2$-PCPP-IZN, bands like 1150, 1060, 530, 3480 cm^{-1} confirm the successful incorporation of rGO-MoS$_2$ in PCPP-IZN polymer (Figure. 8.1.D).

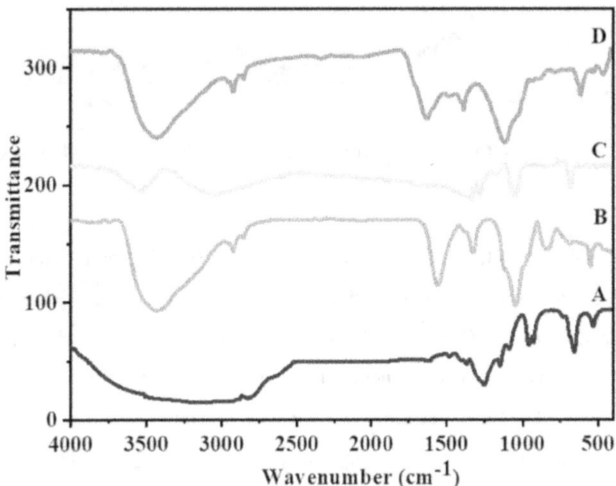

Figure. 8.1. FTIR of (A) GO (B) rGO-MoS₂ (C) PCPP-IZN (D) rGO-MoS₂-PCPP-IZN-ICG.

(ii) XRD pattern

The XRD patterns of GO, rGO-MoS$_2$, PCPP-IZN, and rGO-MoS$_2$-PCPP-IZN were represented in Figure. 8.2. The appearance of (001) diffraction peaks confirm the formation of GO (Figure. 8.2. A). XRD pattern of rGO-MoS$_2$ consist of (002), (001) with higher intensity indicates the reduction of

GO and layered structure of MoS_2. The diffraction peak shows the poor crystallinity of nanomaterials. The (002) plane peak intensity increases show the graphene proposition in rGO-MoS$_2$ (Figure. 8.2. B) [231]. The XRD curve of PCPP-IZN forms a weak, broad peak which shows the amorphous property (Figure. 8.2. C). The incorporation of IZN causes the formation of broad peaks and also a decrease in diffraction peaks. Finally, rGO-MoS$_2$-PCPP-IZN exhibited the diffraction peak (002), (100), (110) which reflection of rGO-MoS$_2$ impregnated in PCPP-IZN (Figure. 8.2. D).

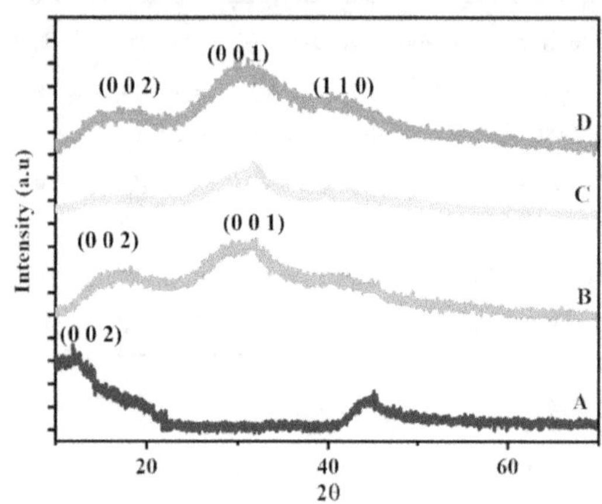

Figure. 8.2. XRD of (A) GO (B) rGO- MoS₂ (C) PCPP-IZN (D) rGO-MoS₂-PCPP-IZN-ICG.

(iii) TGA studies

The thermal stability of GO, rGO-MoS$_2$, PCPP-IZN, and rGO-MoS$_2$-PCPP-IZN was represented in Figure. 8.3. In 100 °C, weight loss was lesser because of the evaporation of surface adsorbed water molecules. GO exhibited major weight loss after 350 °C which is attributed due to the elimination of CO, CO_2, and liable functional groups from GO. Similarly, rGO–MoS$_2$ also exhibited lesser weight loss at 300 °C ascribed to the loss of intercalated water molecules. In higher temperatures, weight loss was corresponding to oxidation of MoS$_2$ to sulfur dioxide and molybdenum oxide. Mostly MoS$_2$ control the oxidation of graphene and subsequent rise in temperature leads to weight loss. PCPP-IZN was less stable to the temperature above 150 °C and further increases in temperature cause residual charring product. The rGO-MoS$_2$-PCPP-IZN exhibited sudden weight loss in initial temperature was due to the polymer degradation. Nanomaterial was stable in higher temperatures because of the availability of rGO-MoS$_2$.

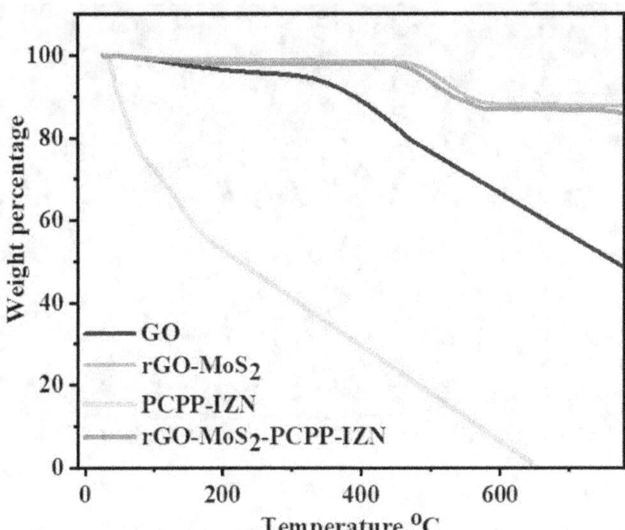

Figure. 8.3. TGA of GO, rGO- MoS₂, PCPP-IZN, rGO-MoS₂-PCPP-IZN-ICG.

(iv) SEM analysis

In Figure. 8.4. (A-C), SEM image deduce the GO ultra-thin, sheet-like layered and it forms a homogenous structure in different range. It forms like a wrinkled paper-like structure and most of the regions are smooth. It also indicates the successful exfoliation of GO into a few-layers. Figure. 8.4. (D-F) illustrates the rGO-MoS₂ interconnected coalescing layers with higher surface area and sheets are stacked together tightly in Figure.8.4 D, E. Figure.8.4. F, shows the formation of 3D sphere-like MoS₂ which were stacked on the graphene layer. The arrangement of MoS₂ in the surface layer enhances the stability of the nanomaterial. The hydrothermal process reduces GO to graphene, and the flexible graphene self-assembled to form 3D architecture [232].

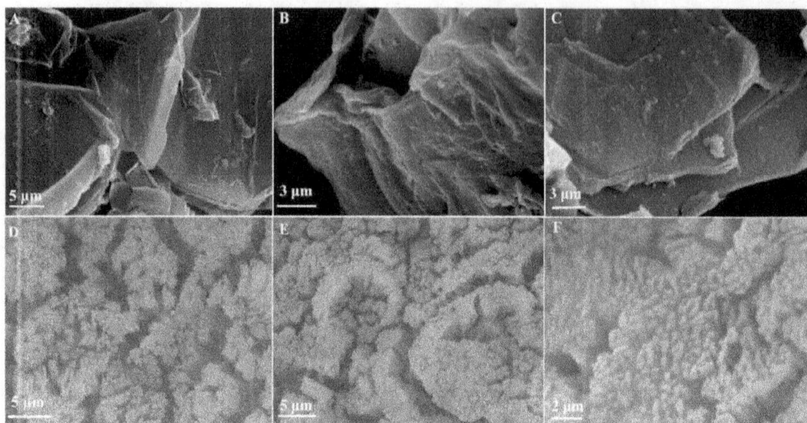

Figure. 8.4. SEM image of GO (A) 5 μm (B, C) 3 μm, rGO-MoS$_2$ (D, E) 5 μm (F) 3 μm.

(v) TEM analysis

In TEM analysis surface studies and morphology of GO were examined which was represent in Figure.8.5. A-F. It confirms the GO formed with more than one-layer, thick flat flake layers structure. It exhibited a smooth surface with a large surface area. In lower magnification, more wrinkled edges were due to the presence of oxygen atoms and a high degree of exfoliation through an oxidation process. The GO forms a semi-transparent material which shows the stability in high-energy beam [233].

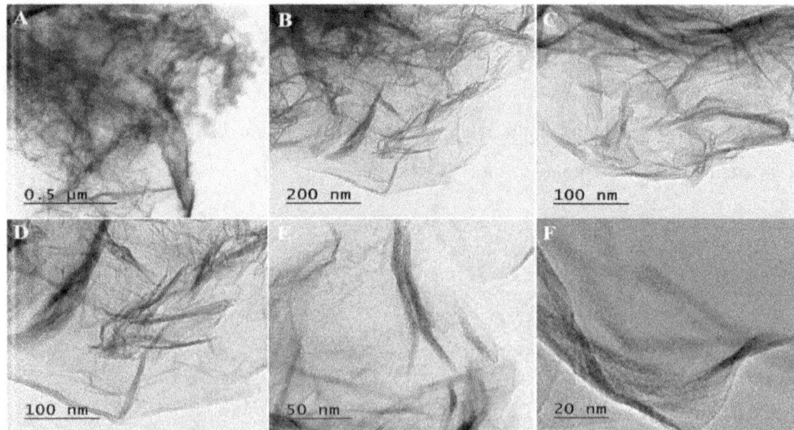

Figure.8.5. TEM image of GO (A) 0.5 μm (B) 200 nm (C, D) 100 nm (E) 50 nm (F) 20 nm.

Through a hydrothermal method, MoS_2 was introduced into graphene to form stable nanomaterials, Figure. 8.6. A-F. It provides visible evidence for the successful formation of rGO-MoS_2. In TEM analysis, it shows the layered platelets, homogenous distribution of MoS_2, and the black stripes indicate MoS_2 contact with rGO. Additionally, surface changes of graphene were also clearly identified which confirms the interactions of MoS_2 and graphene. In different magnification, MoS_2 exhibited the 3D sphere-like structure, randomly distributed on the reduced GO sheets and tightly stacked with the sheet [234]. As with SEM results, the TEM image displays the 3D sphere-like morphology distributed over the sheets. The overlapping 3D structure of rGO-MoS_2 was continued after their annealing. It also forms a layered crystal structure, with crystal lattice fringes with nanosized interlayer [235]. This interlayer space provides a high surface area for the loading of ICG and drug.

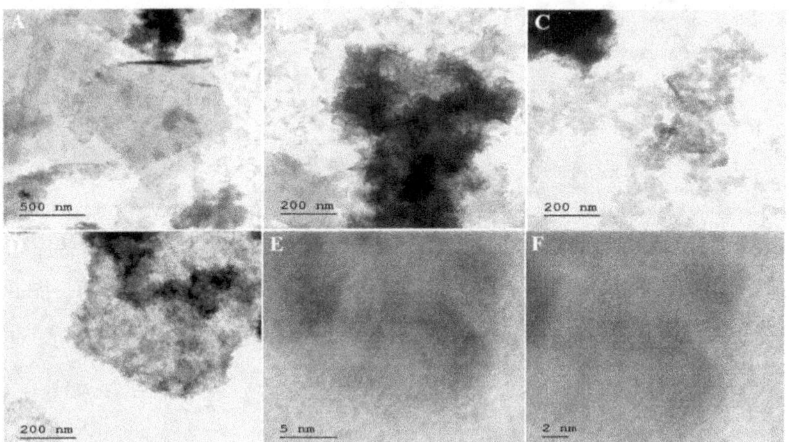

Figure.8.6. TEM image of rGO-MoS₂ (A) 500 nm (B-D) 200 nm (E) 5 nm (F) 2 nm.

(vi) FE-SEM analysis

In Figure. 8.7. (A-D), Field Emission- Scanning Electron microscopy (FE-SEM) of rGO-MoS_2 dopped PCPP-IZN polymer shows nanostructure filling in the polymeric network system. Further, the image confirms the high roughness, which all concludes the presence of rGO-MoS_2 in the polymeric membrane. IZN functionalized PCPP forms a tight packing network to hold the ICG and rGO-MoS_2. FE-SEM image shows the evidence for homogeneous distribution of nanomaterials in the polymeric matrix.

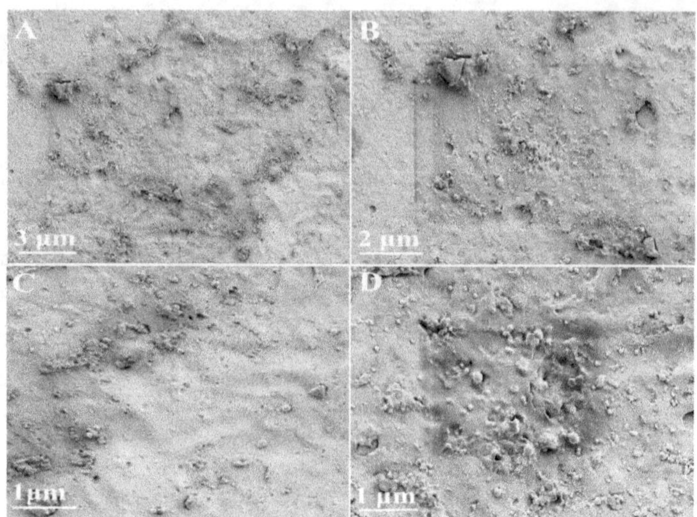

Figure.8.7. FE-SEM image of rGO-MoS₂-PCPP-IZN-ICG (A) 3 µm (B) 2 µm (C, D) 1 µm.

8.6.2. ICG loading

The UV-vis spectra of rGO, rGO-MoS₂, rGO-MoS₂-PCPP-IZN, and rGO-MoS₂-PCPP-IZN-ICG were represented in Figure. 8.8. There is no absorption peak was seemed form for the nanomaterials without ICG in a given range. The rGO-MoS₂-PCPP-IZN-ICG shows the peak shift from 800 to 810 nm which confirms the successful loading of ICG. Nanomaterial-loaded ICG increases the laser light adsorption during the photothermal and photodynamic processe.

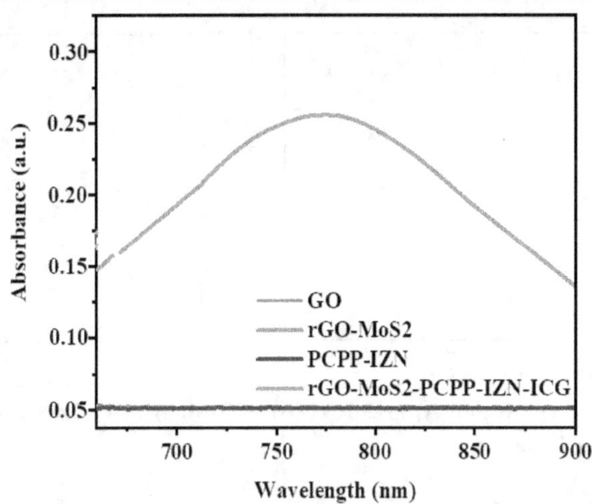

Figure.8.8. The UV-vis spectra of GO, rGO-MoS₂, PCPP-IZN, and rGO-MoS₂-PCPP-IZN-ICG.

Also, for quantitative analysis of ICG loading was represented in Table.8.1, and there is no significant change in ICG loading on rGO, rGO-MoS₂. The maximum ICG loading was 93.34 % on rGO-MoS₂-PCPP-IZN, indicates the high adsorption of ICG by a polymer.

Table. 8. 1. ICG loading efficiency

Sample	rGO	rGO-MoS₂	rGO-MoS₂ -PCPP-IZN
ICG loading	81.3 ± 0.28	84.53±0.43	93.34±0.26

8.6.3. ICG stability

Free ICG was unstable in an aqueous medium, aggregates, and is rapidly degraded to light in a short time. Due to its instability, ICG confines its therapeutic application and its stability improved after conjugation with nanomaterials. The stability of free ICG and rGO-MoS₂-PCPP-IZN-ICG was compared by UV-vis spectra. In Figure. 8.9. A, B. adsorption intensity of free ICG was rapidly decreased in ten days implying the poor stability and maximum degradation of ICG. rGO-MoS₂-PCPP-IZN-ICG exhibited good stabilization which was due to the nanocarrier prevent ICG from aggregation. It also showed 96 % of stability after 10 days which was higher compared to the other rGO [236].

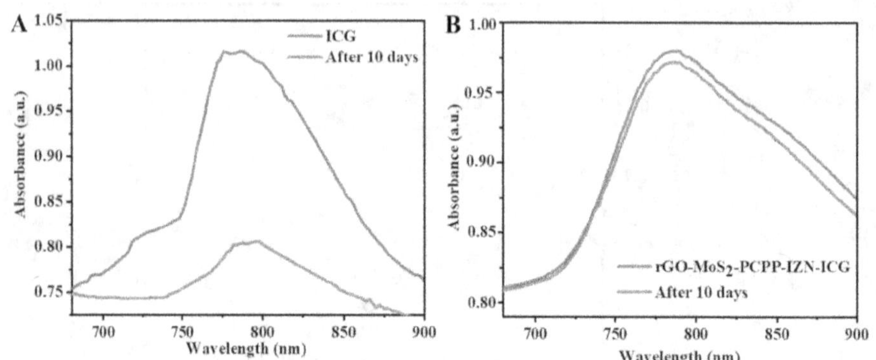

Figure. 8.9. UV-vis spectra of (A) Free ICG (B) rGO-MoS₂-PCPP-IZN-ICG stability compared by measured after 10 days.

8.6.4. *In vitro* IZN and ICG release

In Figure. 8.10. A, The IZN drug release profiles from the rGO-MoS₂-PCPP were evaluated at different pH respectively. Nanomaterial exhibited burst release of drug in lesser time of 120 sec for both pH 5.0 and 7.4. More than 70 % of drug release occurred in 120 min. After 200 sec, IZN follows the cumulative drug release pattern. Compared to the neutral pH, nanomaterials exhibited 83 % of IZN release. This would facilitate higher IZN release in TB infected alveolar macrophage acidic conditions. Further, the ICG release pattern was also evaluated in different pH conditions by UV-vis spectra analysis (Figure. 8.10. B). There were decreases in absorption spectra of rGO-MoS₂-PCPP-IZN-ICG after increasing the pH of the solution. Higher adsorption intensity of ICG loaded nanomaterial upon the pH 4.0 and the intensity was reduced for pH 5.0, 6.0, 7.4. In neutral pH, the adsorption intensity was very low. Electrostatic interaction of the ICG and the nanomaterial ionizable groups results in the pH-based drug delivery.

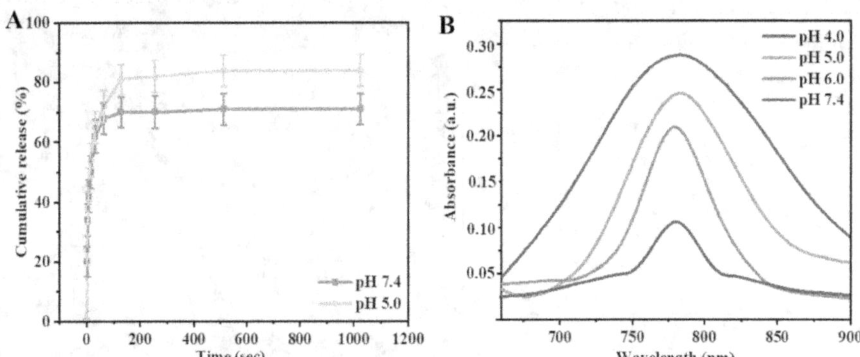

Figure. 8.10. (A) Cumulative IZN release from rGO-MoS$_2$-PCPP-IZN in different pH (B) ICG release from rGO-MoS$_2$-PCPP-IZN-ICG in different pH was measured by UV-vis spectra.

8.6.5. *In vitro* photothermal effect

To evaluate the heat generation ability of ICG, PCPP-IZN, rGO-MoS$_2$, rGO-MoS$_2$-PCPP-IZN-ICG, each nanomaterial suspensions were irradiated with the 808 nm laser in different time interval represent in Figure.8.11. A. PCPP-IZN and rGO has no photothermal effect after treatment with NIR laser. Overall, it records a negligible temperature change of 2 °C. In laser irradiation, the rGO-MoS$_2$ rate of temperature increase was more pronounced within the initial time to 50 °C in 90 sec. The further temperature was increased to 53 °C for 250 sec. The free ICG showed temperature raises from 24 °C to 56 °C in 250 sec, while rGO-MoS$_2$-PCPP-IZN-ICG reached 74 °C. It exhibited a drastic increase in temperature from 25 °C to 74 °C. In 50 sec, it exhibited a temperature of 52 °C, which can destroy the TB bacteria cells. Finally, rGO-MoS$_2$-PCPP-IZN-ICG shows an increase in temperature which was greater than rGO-MoS$_2$ and free ICG. So, this ICG loaded nanomaterial provided a rapid generation of heat abundantly in 250 sec. The structural changes in nanomaterial after irradiation was evaluated and represented in Figure. 8.11. (B-E) Before NIR irradiation rGO-MoS$_2$-PCPP-IZN-ICG exhibited the 3D architecture, tightly stacked nanosheets, sphere-like MoS$_2$ coalescing with rGO. After irradiation rGO-MoS$_2$-ICG structural morphology was completely changed, aggregated, nanomaterial becomes elongated [237]. Most probably the maximum amount of ICG was denatured and it leads to the aggregation of nanomaterials.

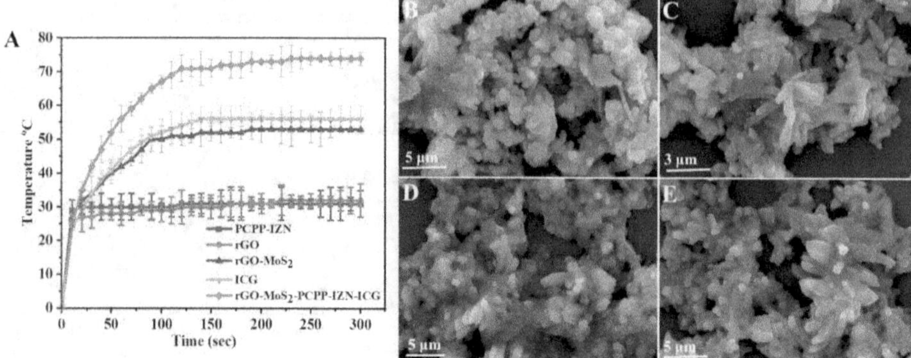

Figure. 8.11. (A) *In vitro* **photothermal effect of PCPP-IZN, rGO, rGO-MoS$_2$, ICG, rGO-MoS$_2$-PCPP-IZN-ICG (B,C,D,E) SEM image of rGO-MoS$_2$-PCPP-IZN-ICG after NIR-irradiation for 5 min.**

8.6.6. *In vitro* PDT studies

ICG produces single oxygen generation ($^1O_2$) which acts as a key factor for PDT. The extent level of $^1O_2$ production was evaluated by absorbance reduction N,N dimethyl-4-nitrosoaniline (DN) using UV-vis spectroscopy. The free ICG, rGO-MoS$_2$-PCPP-IZN-ICG was mixed with DN and irradiated with 808 nm NIR laser for 5 min. The absorbance of the mixture at 404 nm indicates the $^1O_2$ level. NIR laser irradiation of rGO-MoS$_2$-PCPP-IZN and DN mixture shows no change in absorbance at 440 nm before/after irradiation for 5 min (Figure. 8.12. A). It indicates the lack of $^1O_2$ production from the nanomaterial after NIR laser irradiation. But rGO-MoS$_2$-PCPP-IZN-ICG and DN mixture shows a decrease in absorbance after the NIR laser irradiation for 5 min (Figure. 8.12. B). This shows the capability of nanomaterial conjugated ICG in photodynamic activity.

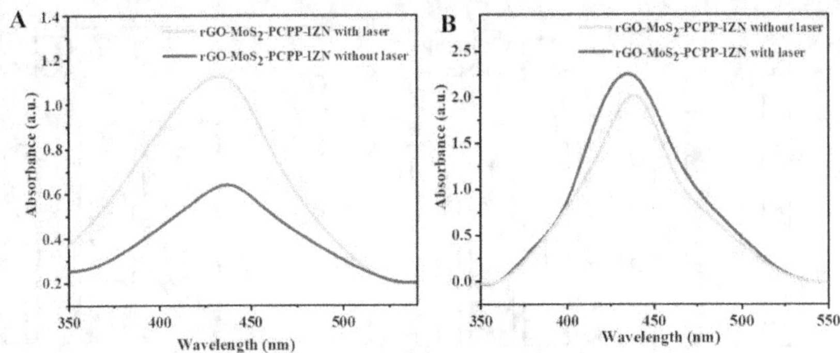

Figure. 8.12. (A) Singlet generation of rGO-MoS₂-PCPP-IZN without and with laser for 5 min (B) Singlet generation of rGO-MoS₂-PCPP-IZN-ICG without and with laser for 5 min.

8.6.7. Antibacterial studies

The antibacterial effect of rGO-MoS₂-PCPP-IZN-ICG nanomaterial after light irradiation for 5 min was examined against the *E.coli* (Figure. 8.13.A) and *B.subtilis* (Figure. 8.13.B). Bacterial culture without any treatment like a laser, photosensitizer was used as controls. There is no significant change in both bacterial cultures after laser light treatment without ICG and nanomaterials. PCPP-IZN reduces the bacterial survival to 37 % (*E.coli*) and 40 % (*B.subtilis*) which was due to the IZN drug. In absence of NIR light, ICG has no antibacterial effect and after irradiation, it reduces bacteria count. Similarly, rGO-MoS₂ shows a good antibacterial effect on both before/after irradiation. This result shows the functionalization of MoS₂ improves the antibacterial effect along with thermal activity [238]. Antibacterial activity of rGO-MoS₂-PCPP-IZN-ICG after NIR irradiation for 5 min shows maximum bacterial cell death. It significantly decreases the *E.coli* and *B.subtilis* bacterial count to 90 %, and 82 %, respectively. This due to the synergistic combination of MoS₂, PCPP-IZN nanomaterials, ICG photothermal, and photodynamic effect.

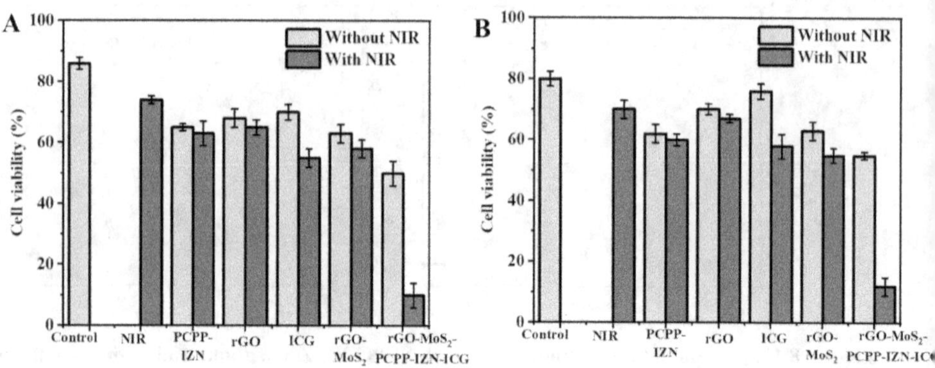

Figure.8.13. Effect of rGO-MoS₂-PCPP-IZN-ICG inhibition in NIR-irradiation on (A) *E.coli* **(B)** *B.subtilis***.**

8.6.8. Bacterial inhibition studies

Further identifying the possible mechanism of antibacterial activity, the irradiated bacterial sample was visualized under a microscope (Figure. 8.14). In the control group, *E.coli* and *B.subtilis* bacterial cells were homogenously dispersed, highly stable cell membrane. Both bacterial cells with ICG exposure and 808 nm laser beam irradiation for 150, 300 sec. Initially, the bacterial cell membrane was shrunken, and the appearance of bacterial aggregates. Further, it was increased in bacterial aggregation by raising the temperature and the photothermal effect enhances the bacterial aggregation. Nanomaterial with lower negative zeta values has strong interaction negative bacterial surface, which resulted in bacteria death [239,240]. The rGO-MoS₂-PCPP-IZN-ICG exhibited the maximum cell wall damage in 150 sec, the whole-cell lost its shape and become a black spherical structure. It indicates the effective antibacterial behavior of rGO-MoS₂-PCPP-IZN-ICG. The active killing of the bacterial membrane was triggered by NIR light and it was mainly carried out in four ways, thermal, single oxygen, MoS₂, and isoniazid-based bacterial cell wall damage. The synergistic combination endowed the excellent antibacterial activity under single wavelength NIR light.

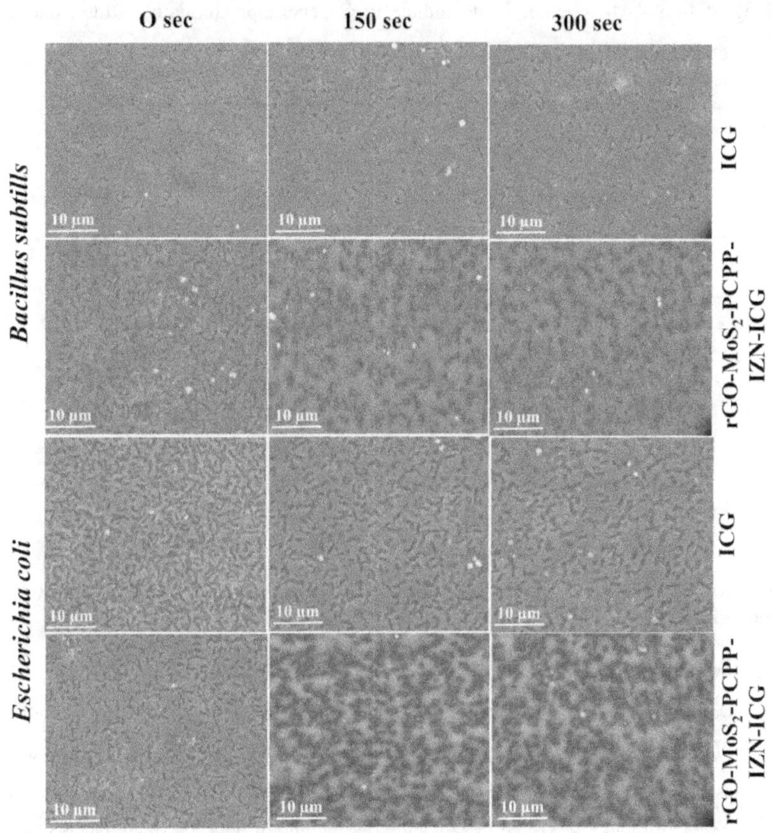

Figure.8.14. Effective photothermal, photodynamic, antibacterial effect of rGO-MoS₂-PCPP-IZN-ICG after NIR-irradiation for different time interval and visualized by phase contrast microscope.

8.6.9. Antimycobacterial activity

In vitro, photothermal and photodynamic effect of rGO-MoS₂-PCPP-IZN-ICG against the *M.tuberculosis* represents in Figure. 8.15. The result shows growth inhibition was based on the concentration-dependent manner. In 2 and 5 mg/mL, bacterial growth was significantly reduced in 200 sec and which was effective compare to the other concentration. NIR radiation causes high photothermal heat inside the *M.tuberculosis*, ICG, and MoS₂ combination results in the generation of single oxygen-lead to maximum cytotoxicity. Also, IZN enhances the antimycobacterial effect by inhibiting of InhA. All together could induce ROS generation, oxidative stress, damages bacterial DNA and membrane integrity [241].

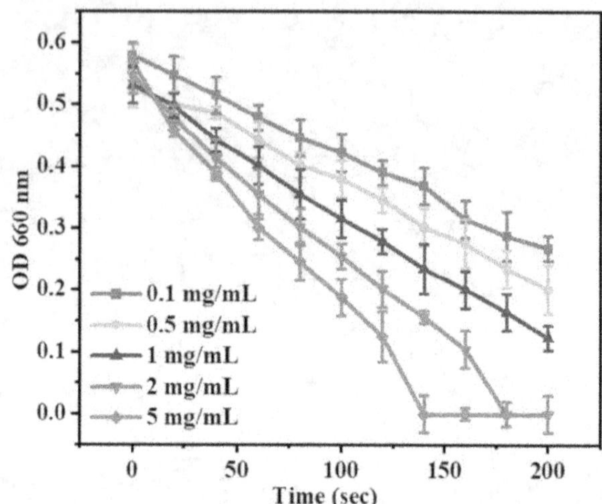

Figure. 8.15. *Mycobacterium* **growth inhibitory effect of rGO-MoS₂-PCPP-IZN-ICG in different concentration.**

8.6.10. Flow cytometry

The antimycobacterial activity was further confirmed by flow cytometry by treating ICG, rGO-MoS₂-PCPP-IZN, and rGO-MoS₂-PCPP-IZN-ICG with *M.tuberculosis* by gating the dead cells. ICG alone has a peak shift and shows the dead cells around 70 % (Figure. 8.16. A). The rGO-MoS₂-PCPP-IZN treatment exhibited 85 % of cell death and after loading of ICG cell death was 98 % (Figure. 8.16. B). The peak shift indicates the maximum cell death due to the synergistic effect of rGO-MoS₂-PCPP-IZN-ICG (Figure. 8.16. C). Therefore rGO-MoS₂-PCPP-IZN-ICG delivers improved novel therapeutic strategies against drug-resistant *M.tuberculosis*.

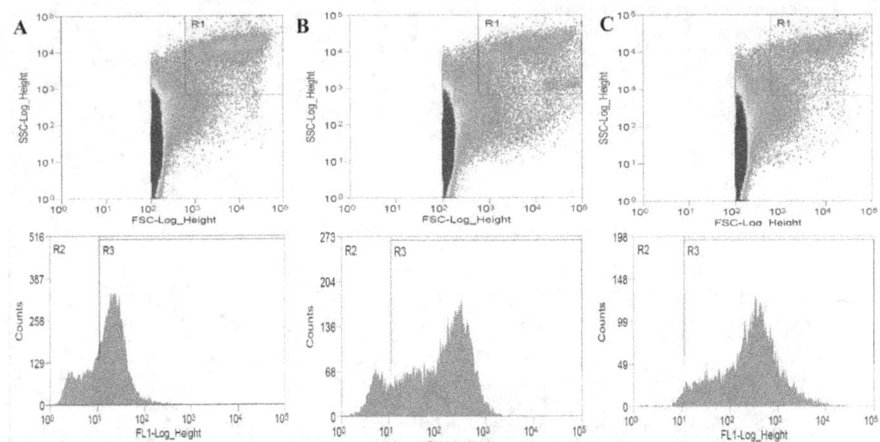

Figure. 8.16. Flow cytometry for identification of viable/dead cells along with histograms peak shift of dead cells (A) ICG (B) rGO-MoS$_2$-PCPP-IZN (C) rGO-MoS$_2$-PCPP-IZN-ICG on *M.tuberculosis*.

8.6.11. SEM analysis

The rGO-MoS$_2$-PCPP-IZN-ICG causes the membrane degradation of *M.tuberculosis* after NIR-irradiation for 5 min. Hyperthermia generated by rGO-MoS$_2$-PCPP-IZN-ICG along with singlet oxygen production and inhibition of InhA causes the damage of cell membrane. It shows higher disruption compared with *M.tuberculosis* without nanomaterial treatment as in Figure. 8.17. (A-C). The larger surface area and layered nanostructures help the higher interaction with *the Mycobacterium* membrane. Majorly, rGO-MoS$_2$ plays an essential role in the binding of the cell membrane, physical cutting, and disruption of membrane integrity (Figure. 8.17. D-F). The combination of PTT, PDT, and chemotherapy leads to the maximum amount of membrane stress by hyperthermia and ROS. So, a combination of all the components damages the bacterial protection three layers and killing them in a shorter timer period of 5 min.

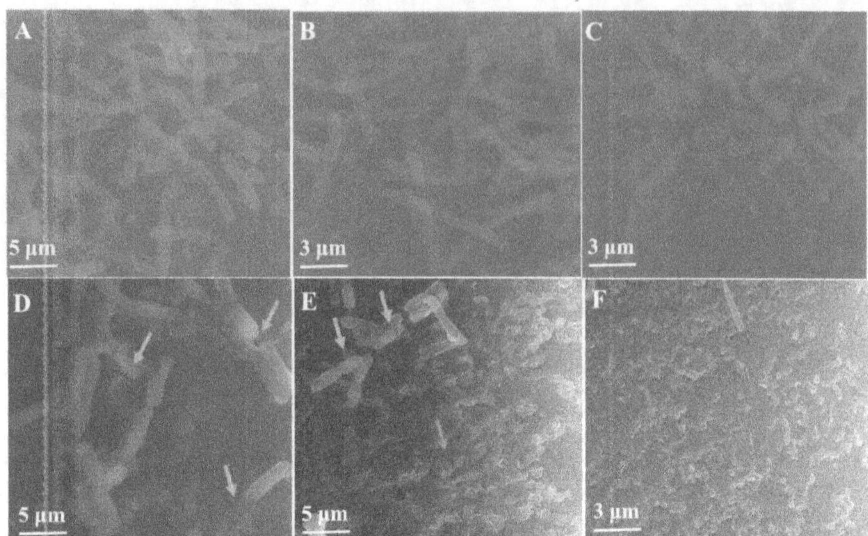

Figure. 8.17. (A-C) SEM images of *M.tuberculosis* without nanomaterial and NIR-irradiation **(D-F)** SEM images of *M.tuberculosis* after nanomaterial treatment and 5 min NIR-irradiation. Blue arrows indicate the cell damage and red arrows show the complete damage of *Mycobacterium* cell.

8.6.12. Cytotoxicity assay

The cytotoxicity of nanomaterial was evaluated on U937 cells by MTT assay, initially different concentration of rGO-MoS₂-PCPP-IZN-ICG was examined (Figure. 8.18. A, B). Nanomaterial shows lesser toxicity to the U937 cells and its IC₅₀ value was 150 µg/mL. The same concentration was used to assess the PCPP-IZN, rGO, rGO-MoS₂, ICG. The toxic nature of rGO-MoS₂ was suppressed by the polymer combination.

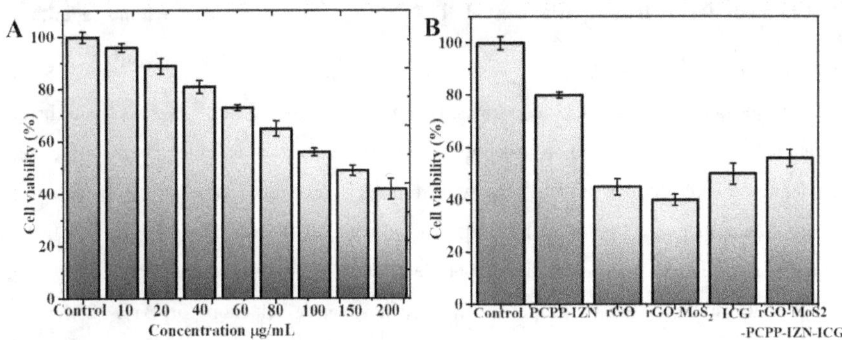

Figure. 8.18. (A) Cytotoxicity of rGO-MoS₂-PCPP-IZN-ICG in different concentration (B) Comparison of cytotoxicity ability of nanomaterials.

8.7. Conclusion

The photoactive nanoagent, rGO-MoS₂-PCPP-IZN-ICG has effective antimycobacterial activity. Concretely, it not only exhibits the photothermal activity under NIR exposure but also works on singlet oxygen generation and antimycobacterial drug release. The bacterial membrane damage starts from the NIR-triggered nanomaterials which were systemically confirmed by *in vitro* studies. The utilization of rGO have a platform for MoS₂ functionalization, carries ICG and IZN drug. The underlying mechanism for bacterial cell membrane damage ability was proposed as follows: (1) Raise of temperature leads to hyperthermia damage of cell membrane, (2) Singlet oxygen enhances the ROS based membrane damage, (3) IZN drug release from nanomaterial stops the mycolic acid biosynthesis by inhibiting the InhA. This process accelerated the death of vulnerable *Mycobacterium* which is achieved by the single nanosystem. Overall, the fabricated all-in-one phototherapeutic nanoagent poses a rational and biosafe technique for eradicating TB diseases.

Bio-mediated hydroxyapatite coated 3D-PCPP polymer scaffolds for bone regeneration

9.1. Introduction

Major bone defects like osteoporosis, trauma, congenital malformations, osteoarthritis provide a great challenge to patients and ortho-surgeons. The currently available therapeutic strategies promote the healing of severe bones defects. Although, limited availability (autograft), poor osteoinductivity, and diseases transmission (allograft) are the reason to increase the demand [242,243]. In BTE, bone polymeric-scaffold material that is biocompatible, porous, biodegradable, mechanically strong, and the material with suitable surface properties will be a superior selection for BTE [244,245]. Poly(ε-caprolactone) (PCL) is an excellent polymer, which is approved by US-FDA and easily alterable properties. The major problem in PCL is insufficient mechanical strength, poor biodegradation ability, lesser bioactivity, and lower cellular response. PCL polymer performance was enhanced by modification of polymeric backbone via the introduction of functional groups, blending with other polymers [246,247]. Especially in BTE, the study on the blending system of PCL and PCPP was used for BTE due to its mechanical and functional properties.

The combination of polymer and ceramics used for hybrid scaffold preparation and reproduce the composition/structural characteristics of native bone. Ceramics chemical composition is similar to bone apatite and it induces bone osteogenesis [248,249]. In ceramics, HAP ($Ca_{10}(PO_4)6(OH)_2$, HAP) is a majorly used bone substituent material for filling bone gaps, scaffold matrix, and a coating of metallic implants. Due to its poor mechanical ability, the researcher shifted towards the nano-based technique like sol-gel, wet precipitation, hydrothermal, microemulsion, microwave irradiation method for HAP preparation [250]. Recently, natural sources materials like, eggshells, crab shells, corals, snailshells, shrimp shells, fish-bone, were used for HAP synthesis. This biological origin helps in the recycling of biowaste, provides fine-scale microstructure, alterable crystalline property, porosity, morphology, and notable mechanical property. Among the several materials, snail shells are abundant in nature and carry calcium carbonate, trace elements, and organic matter [251-253].

9.2. Previous work

Only, few studies were utilized the snail shells for HAP synthesis, snail shell powder subjected to high-temperature calcination to form calcium oxide (calcium source) [254,255]. Suresh Kumar *et al.*, have synthesized the HAP from snail shells by hydrothermal method and used the phosphorus source as an (NH_4) $2HPO_4$ [256]. The blending of PCL and PCPP is rare. It is an easy fabrication into desired shapes, optimize the biodegradation rate and tunable mechanical properties for a specific application. Several studies show the scaffold formation ability of PCPP, 3D polymer structure alone promote the osteoblast adhesion and proliferation.

9.3. Novelty

Three different types of shells like Tower snail shell (TSS), Bicoloured Fan shell (BFS), Clamshell (CS) were used for HAP synthesis. These snail shells were a good source of calcium carbonate and trace elements Sr, Zn, Al, P, K, Ca, Na, Fe, V, Cd, and Pb. However, the bioactive ceramics is limited due to poor scaffold forming ability, their fragility, and low biodegradation rate [257]. Also, polymeric scaffolds are poor in retaining their original shape for a longer time [258,259]. So, the development of an interconnected polymer and bioceramics hybrid scaffold have gained both component advantages.

9.4. Objective

The aim was to design 3D porous PCL-PCPP based scaffolds coated with HAP for multiple functionalities like improved mechanical properties, enhanced biomineralization, and bioactive component delivery. The hybrid polymeric scaffolds were prepared by solvent casting-particulate leaching method. It forms a scaffold with interconnected, highly porous material with desired pore size which was characterized FTIR, XRD, and SEM. Further, snail shell HAP grafted scaffold biomineralization, osteoblast cell adhesion, and regeneration ability were identified using MG-63 cells.

9.5. Experimental techniques
9.5.1. Preparation of PCPP/PCL porous scaffold

The PCPP, PCL of different weight ratios was used to prepare the porous polymeric scaffold via the conventional solvent-casting/particulate-leaching method [260]. Briefly, PCPP, PCL was dissolved in 5 mL of chloroform for 1 h at room temperature to form a stable suspension. Then, the polymeric solution was mixed with NaCl particulate, cast onto the molds (diameter-10 mm) and chloroform was left to evaporation for 24 h. In meantime, polymeric composites were manually pressed for every 1 h to remove bubbles. Later, the composites were removed from the molds and immersed into deionized water for 2 days to leach out the NaCl particulate. After 48 h, the scaffold was dried under vacuum-oven and scaffold intended with a weight ratio of PCPP: PCL-1:1,2:1 and 3:1 was labeled as PCPP1/PCL, PCPP2/PCL, and PCPP3/PCL scaffold.

9.5.2. Synthesis of HAP from snail shell

Different types of shells like *Turritella duplicata* (Tower snail shell-TSS), *Pinna bicolor* (Bicoloured Fan shell-BFS), *Mercenaria mercenaria* (Clam shell-CS) were collected and rinsed in tap water, followed by distilled water. The washed shells were grounded into a fine powder and mixed in sodium hypochlorite to complete the removal of organic components. Then, washed with water several times and dried at 120 °C for 5 h. Dried shell powder was calcined at 1000 °C for 2 h to form calcite and three different HAP was synthesized separately. Approximately, 1 g of respective shell powders was added to 0.6 M H_3PO_4 solution with continuous stirring and pH 10 was maintained with ammonium hydroxide solution. Further, the precipitate was washed, filtered, and dried at 100 °C for 4 h. Further, the fine powder was sintered at 1000 °C for 4 h to form HAP and it labeled TSS-HAP, BFS-HAP, and CS-HAP based on the common name of shell.

9.5.3. Fabrication of HAP coated PCPP3/PCL scaffold

TSS-HAP was coated on the PCPP3/PCL scaffold via dip-coating technique [261]. Different concentration of shell HAP suspension was formed by mixing-0.5, 1, 1.5 and 2 % of TSS-HAP in 10 mL ethanol and it maintained in bath sonication for 2 h. The PCPP3/PCL scaffold was dipped in TSS-HAP suspension for 3 min and it was taken out for air drying. This process was repeated 3 times and TSS-HAP coated scaffold was labeled as TSS-HAP1/PCPP3/PCL, TSS-HAP2/PCPP3/PCL, TSS-HAP3/PCPP3/PCL, and TSS-HAP4/PCPP3/PCL scaffold based on TSS-HAP concertation 0.5, 1, 1.5 and 2 %.

9.5.4. Scaffold porosity measurement

The scaffold's porous nature was determined by the liquid displacement method [262]. The scaffold was completely immersed in hexane of known volume (V1) for 3 h and the hexane-impregnated scaffold volume was recorded (V2). The scaffold along with the hexane in its pores was removed and the residual liquid volume was measured (V3). The porosity (%) was calculated by the following equation,

$$\text{Porosity \%} = \frac{(V1 - V3)}{(V2 - V3)} \times 100 \text{ \%}$$

9.6. Results and discussion

In this work, a porous PCL/PCPP scaffold was prepared by the salt leaching method and it forms a 3D-polymer scaffold with the desired shape. In this method, pores are formed by porogen (NaCl), which can simply be cleared from the scaffold by washing with water. The porous 3D scaffold was coated with natural shell-based HAP by the dip-coating method. HAP was incorporated to facilitate the mineralization, serve as an anchor site for osteoblast cells, and regeneration. The functionalization of HAP, structural, porosity, mechanical properties of scaffold, and osteoblast proliferation studies was analyzed.

9.6.1. Characterization of PCPP/PCL scaffold
(i) FTIR and XRD analysis

To examine the components of the scaffold and confirm the PCPP/PCL interaction in various concentrations was analyzed by FTIR analysis (Figure. 9.1. A, B, C). The bands around 1730 cm^{-1} represent the PCL carbonyl stretching and 3488 cm^{-1} indicate the water absorption. The peak around 2986, 2876 cm^{-1} represents the vibration of $-CH_2$ group, and 1743 cm^{-1} denotes the PCPP carboxylic acid C=O peaks. In the XRD image Figure. 9.1. D, E, F, have a major reflection of the sharp peak at 21, 23° represents the semicrystalline PCL and the attained results are in accordance with the literature. The change in the intensity of distinct peaks is based on the PCPP concentration.

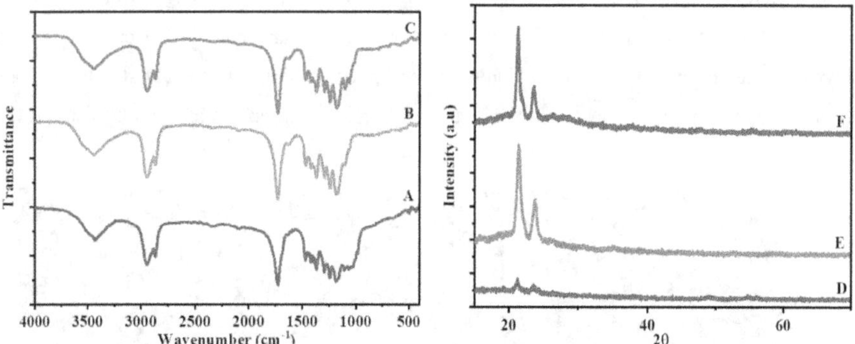

Figure. 9.1. FTIR, XRD image of (A, D) PCPP1/PCL scaffold, (B, E) PCPP2/PCL scaffold, (C, F) PCPP3/PCL scaffold.

(ii) Porosity and mechanical strength of scaffold

The porosity of the PCPP/PCL scaffold was 87-88 % with high interconnected pore networks (Table. 9.1). This enhanced pore morphology in the scaffold was vital for the transport of the cells, metabolites, blood circulation, and vascularization [263]. In 1 % PCPP added scaffold easily failed under compressive load due to its brittleness. The scaffold with 3 % PCPP showed Young's modulus of 34.12 ± 3.19 MPa, compressive yield strength: 1.46 ± 0.07 MPa. Higher compressive ability was due to the PCPP superior strength, which provides dense structure, and tight connection in the scaffold [264].

Table. 9.1. Porosity and mechanical strength value of different concentration of PCPP/PCL scaffold.

PCL (%)	PCPP (%)	Young's modulus (MPa)	compressive yield (MPa)	Porosity
1	1	8.64± 3.25	0.24± 0.53	87.01 ± 2.66
1	2	14.32± 2.02	1.0± 0.46	87.44 ± 1.07
1	3	34.12 ± 3.19	1.46 ± 0.07	88.26 ± 2.46

(iii) SEM analysis

Scaffold prepared using solvent-casting/particulate-leaching from an interconnected porous scaffold with excellent mechanical strength. Figure. 9.2. A-F, it shows homogeneous regular morphology with pores distributed throughout the scaffold architecture. The pores in the PCPP3/PCL scaffold were strongly connected and interlinked network. The porosity ranged from 87-88 %, which was related to the NaCl particulate size. The surface texture and pore wall were rougher, stronger polymer skeleton without any cracks indicates the higher stability of scaffold structure. This dense, intact porous structure enhances the load-bearing strength and osteoblastic cell adhesion, proliferation [265].

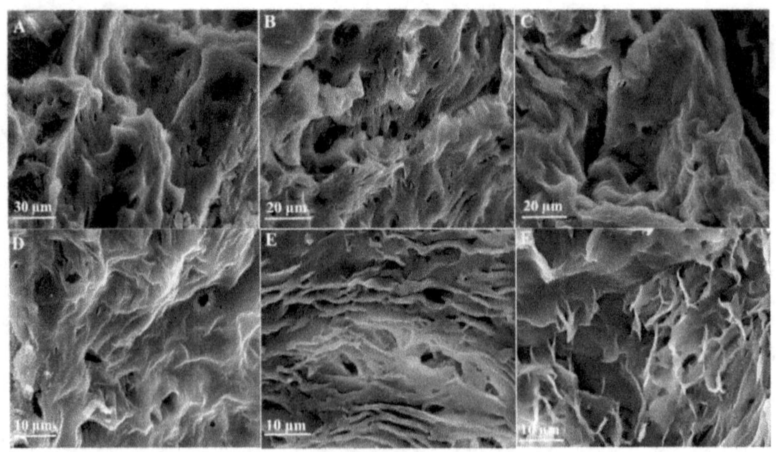

Figure. 9.2. SEM image of PCPP3/PCL scaffold in various magnification (A) 30 μm (B, C) 20 μm (D, E, F) 10 μm.

9.6.2. Characterization of snail shell HAP
(i) FTIR analysis

HAP was synthesized using calcite of *Turritella duplicata* (TSS), *Pinna bicolor* (BFS), *Mercenaria mercenaria* (CS) via the chemical precipitation method. This calcite was decomposing into CaO, and CO_2, then NH_4HPO_4 forms an NH_4OH, and PO_4^{3-} at a constant temperature with raising the mixing time. The FTIR spectrum of three different HAP was compared, which shows the PO_4^{3-} and hydroxyl groups vibration represents in Figure.9.3. A, B, C. Some major vibrational peak at 546, 602, 1004 cm^{-1} indicates the tetrahedral phosphate group of HAP. The hydroxyl peak stretching vibration at 3514 cm^{-1} and additional peak at 1414 cm^{-1} is due to the CO_3^{2-} which present only in BFS-HAP and CS-HAP. In TSS-HAP, CO_3^{2-} the peak was disappeared which almost not exist and PO_4^{3-} vibration peak intensity was higher compared to the BFS and CS HAP.

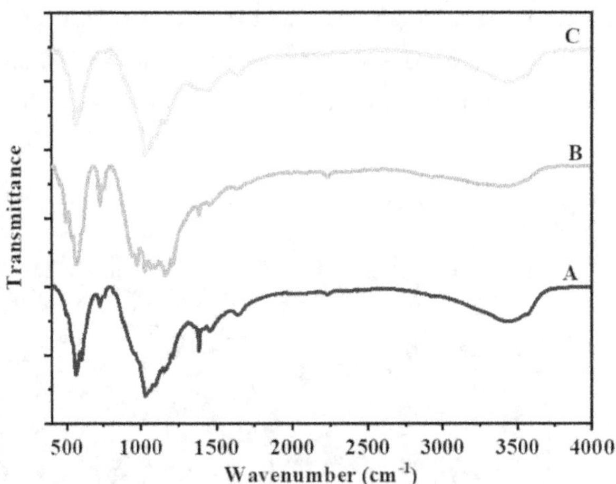

Figure. 9.3. FTIR spectra of (A) HAP from *Turritella duplicate* (TSS-HAP), (B) HAP from *Pinna bicolor* (BFS-HAP) (C) HAP from *Mercenaria mercenaria* (CS-HAP).

(ii) XRD pattern

The XRD pattern of three HAP was represented in Figure. 9.4. A, B, C, which was compared with HAP JCPDS data (file no. 09-0432). In that BFS-HAP and CS-HAP shows the major phases of calcite and TSS-HAP represents the more similarity to the HAP JCPDS data. It indicates the pure crystalline HAP phase with a hexagonal crystal system.

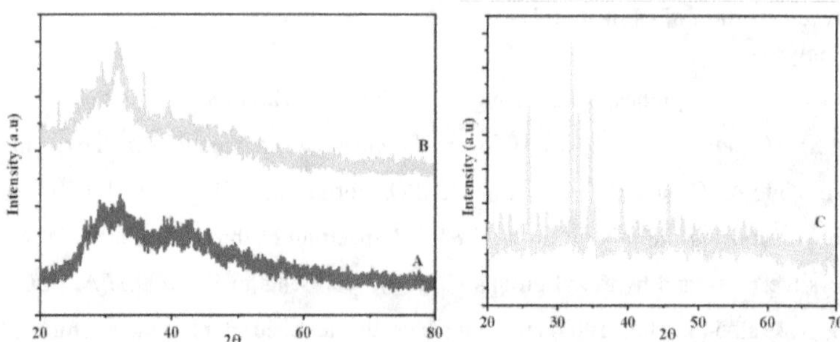

Figure. 9.4. XRD spectra of (A) HAP from *Turritella duplicate* (TSS-HAP), (B) HAP from *Pinna bicolor* (BFS-HAP), (C) HAP from *Mercenaria mercenaria* (CS-HAP).

In SEM observation, TSS-HAP forms fine particles, the morphology of HAP is non-spherical, and angular structure (Figure. 9.5. A, B). The EDAX of TSS-HAP was taken from the induvial particles, which confirms it composed of calcium, phosphorous, oxygen, strontium, zinc, magnesium, and silica (Figure. 9.5. C). These ions diffuse the surface and locate at the lattice sites of crystalline structure.

(iii) SEM and EDAX of TSS-HAP

Figure. 9.5. (A,B) SEM image of HAP from *Turritella duplicate* (TSS-HAP) (C) EDAX analysis of HAP from *Turritella duplicate* (TSS-HAP).

9.6.3. Characterization of TSS-HAP coated PCPP3/PCL scaffold
(i) FTIR analysis

The TSS-HAP was used for coating the PCPP3/PCL scaffold via the dip-coating method and initially, the scaffold was coated via different concentrations of TSS-HAP 1-4 (0.5, 1, 1.5, and 2 %). To determine the effect of HAP concentration on the fabrication of a better HAP scaffold. Figure.9.6. A-D, shows the characteristics band of TSS-HAP coated PCPP3/PCL scaffold, bands 1730, 2986, 2876 represent the carbonyl stretching, $-CH_2$ group, C=O peaks of PCPP3/PLA. The formation of stretching bands at 565, 606, 966, and 1033 cm^{-1} represents the vibration of PO_4^{3-}. Additionally, the PO_4^{3-} vibration peaks were increases with TSS-HAP concentration.

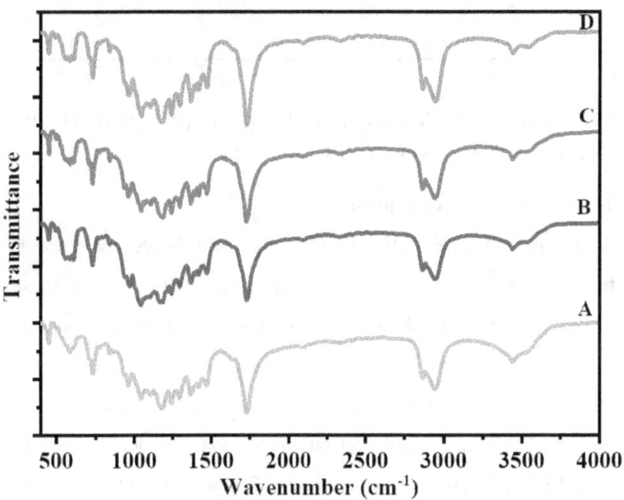

Figure. 9.6. FTIR of (A) TSS-HAP1PCPP3/PCL (B) TSS-HAP2PCPP3/PCL (C) TSS-HAP3PCPP3/PCL (D) TSS-HAP4PCPP3/PCL scaffold.

(ii) XRD pattern

The phase change of PCPP3/PCL scaffold due to TSS-HAP coating was analyzed by XRD (Figure. 9.7. A-D). The small hump, border peak at 22° in all the samples corresponds to the PCPP3/PCL interaction. Further, the sharp peaks at 31, 34° corresponds to the crystalline TSS-HAP presence in the PCPP3/PLA scaffold, and also, the intensity increases with the TSS-HAP concentration.

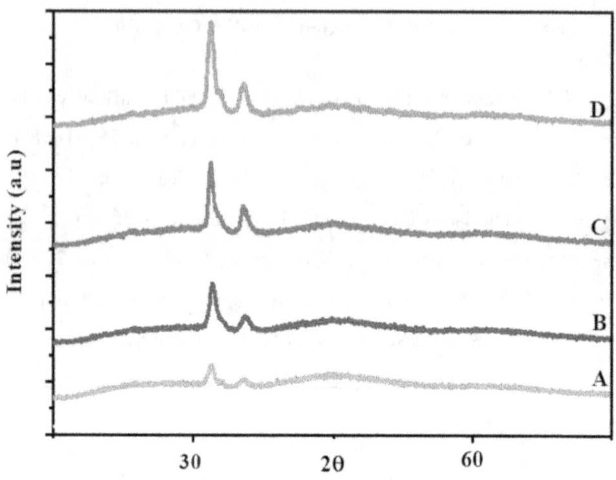

Figure.9.7. XRD image of (A) TSS-HAP1PCPP3/PCL (B) TSS-HAP2PCPP3/PCL (C) TSS-HAP3PCPP3/PCL (D) TSS-HAP4PCPP3/PCL scaffold.

(iii) SEM and Elemental mapping analysis

In Figure. 9.8. A-D, TSS-HAP4/PCPP3/PCL scaffold shows the compact, regular morphology, homogenous structure, strongly interlinked, and open-pore structure which is vital for bone tissue regeneration. In Figure. 9.8. E-H, TSS-HAP1/PCPP3/PCL shows the non-uniform distribution of HAP, dispersed in the lesser region of the scaffold, and agglomerated. The uneven distribution of HAP leads to the alteration of the scaffold integrity, pore size, and structure. PCPP3/PCL scaffolds were simply immersed with HAP and it forms an interaction via Van der Waals forces or hydrogen bond onto the surface [266]. So, its appearance and interaction were based on the optimum concertation of HAP. In 2 % of HAP coating on PCPP/PCL scaffold, shows the even distribution of spherical like structure HAP on the scaffold surface and it was optimum concentration for coating.

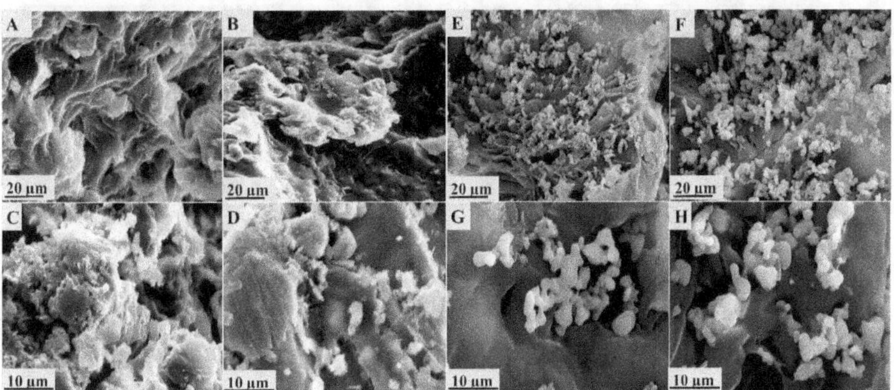

Figure. 9.8. SEM image of (A-D) TSS-HAP1PCPP3/PCL scaffold in different magnification (E-H) TSS-HAP4PCPP3/PCL scaffold in different magnification.

The elemental mapping images of the TSS-HAP4/PCPP3/PCL scaffold were performed and the elements like Sr, Zn, Mg, Si, Ca, and P were seen in the Figure. 9.9. The red (Sr), cyan (Ca), violet (Si), yellow (P) colors spread over the scaffold surface. The green (Zn), blue (Mg) color was majorly spotted in the corner and it denotes the lesser dispersed on the scaffold. The presence of the trace elements like Sr, Si, Zn, Mg on the scaffold is derived from the tower snail shell. These trace minerals are essential for bone growth by playing a major role in stimulates osteoblast activity, proliferation, bone remodeling, inhibition of osteoclast, and mineralization [267].

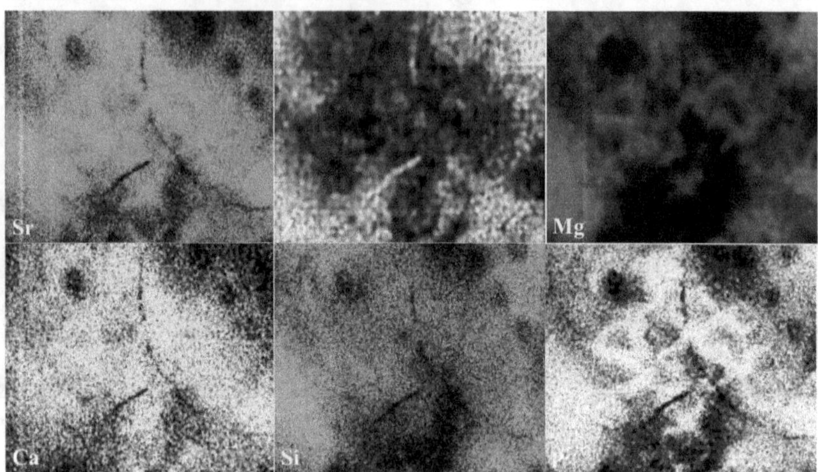

Figure. 9.9. Elemental mapping of TSS-HAP4/PCPP3/PCL scaffold.

9.6.4. Porosity, mechanical properties of scaffold

Another essential criterion for a scaffold is porosity and it is important to evaluate the pore structure of the scaffold. The different concentrations of TSS-HAP coated PCPP3/PCL scaffold porosity were listed in Table. 9.2. and it showed porosity of almost 76-84 %. The increase of HAP concentration during coating causes decreases in the porosity of the scaffold. In concentration, TSS-HAP1 coating shows the porosity of scaffold was 84 % and which was decrease into 74 % for the TSS-HAP4 coating. In higher concentrations, more amount of HAP was distributed to the scaffolds. The mechanical properties of HAP coated scaffold were represented in Table.9.2 and the compressive strength of scaffold material related to the HAP concentration. The PCPP3/PCL scaffold with 2 % TSS-HAP coating has the highest compressive properties of (42.24±1.23 MPa- Young's modulus, 2.18±1.24 MPa) and it was reduced for 0.5 % TSS-HAP coating. HAP forms strong ionic interaction to scaffold via Ca^{2+} and COO^- of PCCP polymer. A similar result was reported by Milovac *et al.*, which shows an increase in compressive strength of scaffold material based on the HAP coating concentration [268].

Table. 9.2. Porosity, mechanical properties TSS-HAP 1-4 coated PCPP3/PCL scaffold.

Coating	Young's modulus (MPa)	Compressive yield (MPa)	Porosity
TSS-HAP1	38.43 ± 2.28	1.48 ± 1.16	84.18±3.21
TSS-HAP2	39.24 ± 1.48	1.52 ± 1.67	82.26±1.14
TSS-HAP3	40.16 ± 1.39	2.01 ± 1.33	78.18±2.18
TSS-HAP4	42.24±1.23	2.18±1.24	76.18±4.21

9.6.5. TGA studies

TGA study was performed to evaluate the thermal stability of PCPP3/PCL scaffold based on the TSS-HAP concentration. In Figure. 9.10, TG curves illustrate higher thermal stability of TSS-HAP and the weight loss of 5 % at 600 °C. But, the PCPP3/PCL scaffold exhibited 50 % weight loss at 250 °C and it also degrades in two steps. Initial, weight loss was due to the evaporation of water molecules at a temperature above 100 °C. The polymeric decomposition occurred in the second step of weight loss at a temperature range of 500 °C. Polymer consist of more amount of functional group with labile oxygen, which easily forms CO, CO_2 at this temperature. The polymeric scaffold thermal stability was increased by varying the HAP concentration (0.5-2 %). More amount of TSS-HAP treatment with PCPP3/PCL scaffold converts the labile oxygen into the stable functional group and enhances the thermal stability [269]. The final residual weight of scaffolds was 3.1, 6.3, 9.2, and 16.4 % for different concentrations of TSS-HAP.

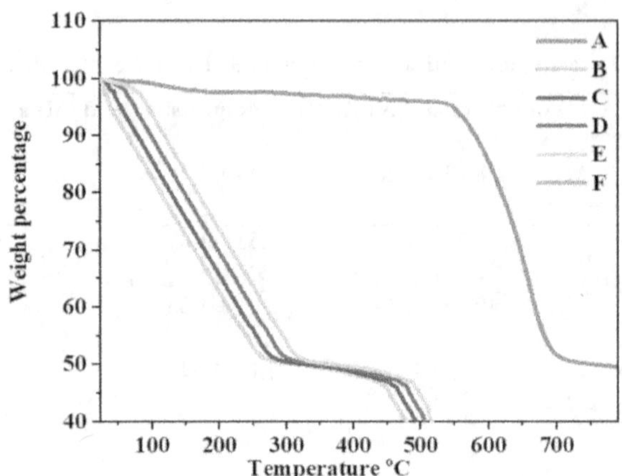

Figure. 9.10. TGA analysis of (A) TSS-HAP (B) PCPP3/PCL scaffold (C) TSS-HAP1PCPP3/PCL scaffold (D) TSS-HAP2PCPP3/PCL scaffold (E)TSS-HAP3PCPP3/PCL scaffold (F)TSS-HAP4 PCPP3/PCL scaffold.

9.6.6. Weight loss of TSS-HAP-PCPP3/PCL scaffold

In vitro weight loss of TSS-HAP coated scaffold material after soaking in PBS for several days was evaluated and shown in Figure. 9.11. The degradation of all the scaffold was based on the incubation time, increases in treatment time cause higher weight loss of scaffold materials. A more amount of degradation occurred in the TSS-HAP1/PCPP3/PCL scaffold. Scaffold coated with TSS-HAP1, TSS-HAP2, TSS-HAP3, TSS-HAP4 were losses weight about 8.6, 7.0, 6.0, 4.6 %, respectively, after 60 days. The PCPP3/PCL scaffold degradation was controlled by TSS-HAP, which delay the circulation of buffer solution into the scaffold. So, this causes the slower degradation of scaffold based on the TSS-HAP concentration.

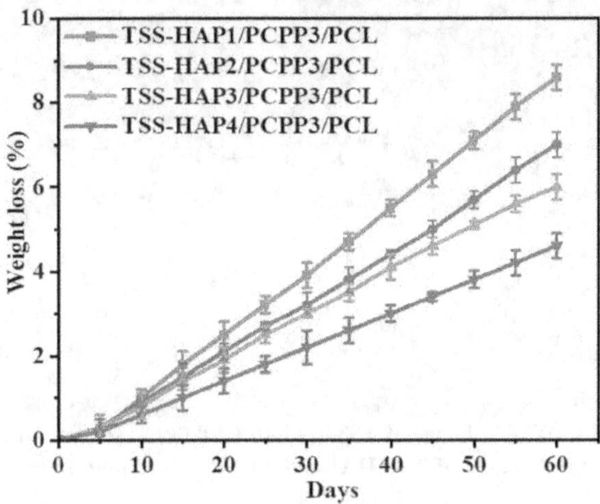

Figure. 9.11. Weight loss of TSS-HAP1, TSS-HAP2, TSS-HAP3, TSS-HAP4 coated PCPP3/PCL scaffold.

9.6.7. Biomineralization test

The level of apatite formation on the TSS-HAP4/PCPP3/PCL scaffold after treatment with SBF for 21 days was visualized by SEM and scaffold surface structural changes were represent in Figure. 9.12. In 7 days of SBF treatment, there is the formation of a spherical apatite-like mineral layer over the scaffold surface. In the presence of SBF, TSS-HAP on the scaffold induces the nucleation of calcium phosphate to form apatite in globular structure (Figure. 9.12. A-D). The apatite layer was flat on the surface of the 3D scaffold and there is the continuous growth of crystal apatite on the scaffold. It was observed in SEM at 21st days, formation of secondary nucleation on the initial mineral apatite and which completely spreads over the scaffold (Figure. 9.12. E-H). It completely modifies the scaffold topography compared to 7 days treatment. Mineral deposited apatite was strongly bound with polymer scaffold induces the growth and adhesion of osteoblast cells [270].

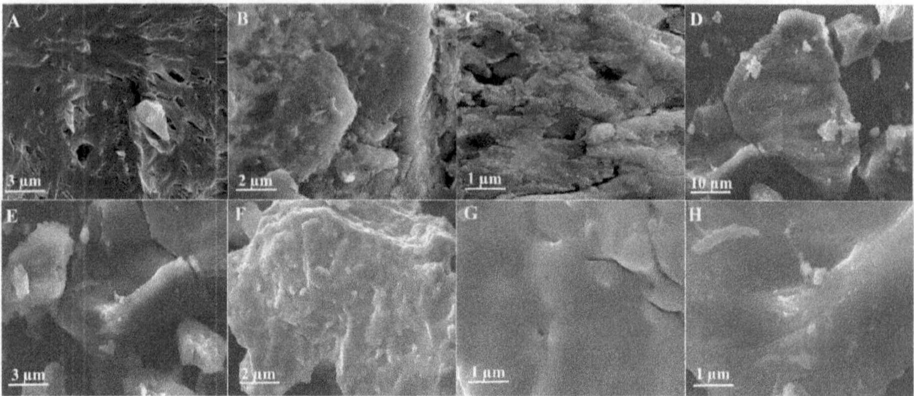

Figure. 9.12. (A-D) SEM image of TSS-HAP4/PCPP3/PCL scaffold after treatment with SBF for 7 days (E-H) SEM image of TSS-HAP4/PCPP3/PCL scaffold after treatment with SBF for 21 days.

9.6.8. Cell proliferation and viability studies

Different concentrations (10 to 60 μg/mL) of TSS-HAP1, TSS-HAP2, TSS-HAP3, TSS-HAP4 coated PCPP3/PCL scaffold was treated with MG-63 cells and evaluated for cell proliferation via MTT (Figure. 9.13). There is an insignificant decrease of cell proliferation for the PCPP3/PCL scaffold of 60 μg/mL concentration. But, MG-63 osteoblast cells displayed higher cell viability and increases in proliferation on all the scaffolds coated with TSS-HAP. Majorly it shows the dose-dependent proliferation of osteoblasts cells and especially, higher proliferation was showed in the 60 μg/mL concentration of TSS-HAP4/PCPP3/PCL scaffold. The maximum proliferation of MG-63 cells was due to the osteogenic induction of TSS-HAP.

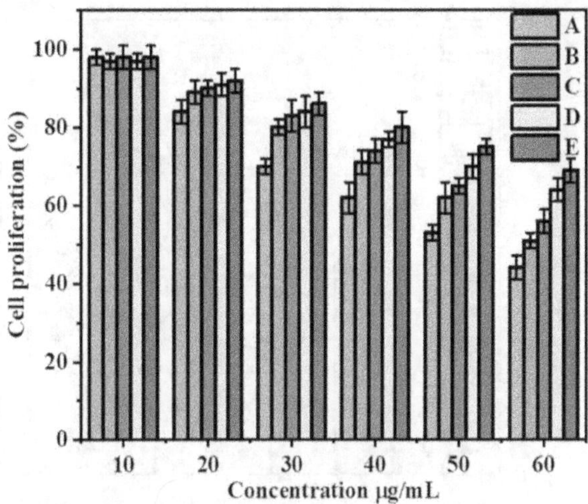

Figure. 9.13. MG-63 cell proliferation rate after treatment with different concentration of (A) TSS-HAP (B) TSS-HAP1PCPP3/PCL (C) TSS-HAP2PCPP3/PCL (D) TSS-HAP3PCPP3/PCL (E) TSS-HAP4PCPP3/PCL scaffold.

9.6.9. ALP activity of scaffold

To evaluate the TSS-HAP coated scaffold potential on osteoblastic activity, a scaffold was treated with MG-63 cells to measures the alkaline phosphatase level. It was performed by p-nitrophenol and scaffold treatment was executed for different time intervals represent in Figure.9.14. There is a steady increase of the ALP activity for TSS-HAP 1 to 4 coated PCPP3/PCL scaffold treatment, from day 7 to 14 and it was higher compared to the PCPP3/PCL scaffold. This highlights the TSS-HAP4/PCPP3/PCL scaffold has higher ALP activity which enhances the osteogenic process, cell differentiation, and colonization.

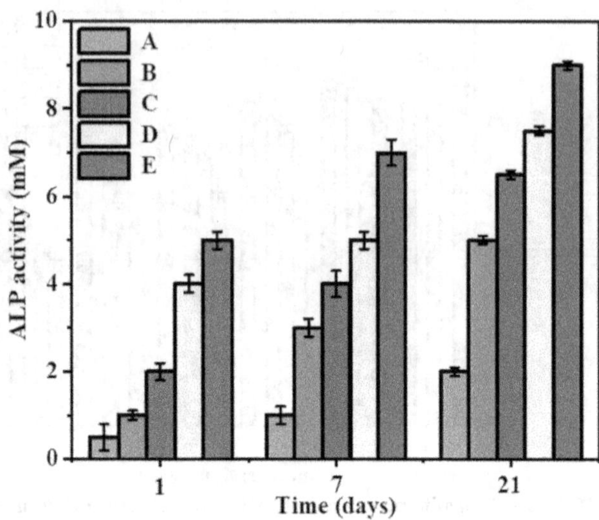

Figure. 9.14. MG-63 cell ALP activity after treatment with different concentration of (A) TSS-HAP (B) TSS-HAP1PCPP3/PCL (C) TSS-HAP2PCPP3/PCL (D) TSS-HAP3PCPP3/PCL (E) TSS-HAP4PCPP3/PCL scaffold.

9.6.10. Clonogenic assay of scaffold

The Clonogenic assay was performed as section. 2.9.3. MG-63 cell's ability for colonies formation from single cells after treatment with TSS-HAP4 coated PCPP3/PCL scaffold for 21 days was measured by clonogenic assay (Figure. 9.15). MG-63 cells without any treatment were examined similarly for 7, 14, and 21 days. The rate of colony formation was very low compared to TSS-HAP4 coated scaffold and which exhibited the high dense colonies on the 7th day. The colonies formation was increased by raising treatment days for 14, and 21. On the 14th day, colonies of cells were dispersed over the well and it exhibited a more blue color. This blue color intensity was maximum on the 21st day and denotes the formation of larger colonies of MG-63 cells. So, it shows the scaffold's ability to trigger the indefinite osteoblast cell proliferation and is also able to form larger colonies from single cells [271].

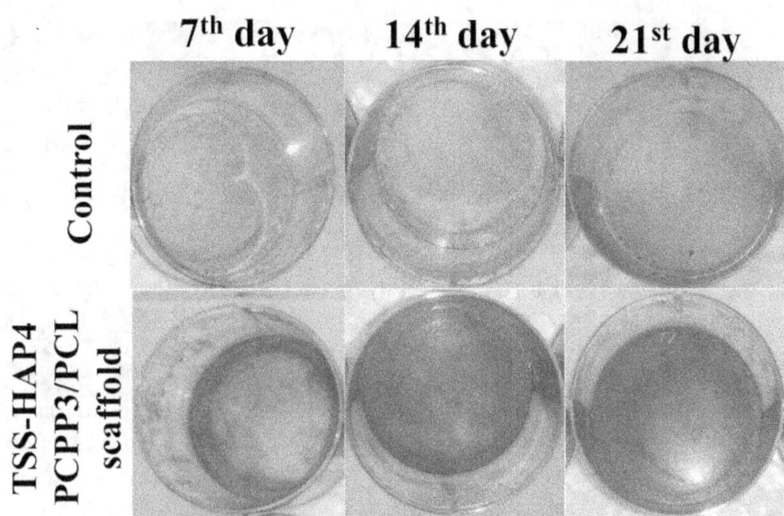

Figure. 9.15. Clonogeny study of MG-63 cells after treatment with TSS-HAP4 coated PCPP3/PCL scaffold for 7, 14, 21 days.

9.6.11. MG-63 cell morophological studies

Additionally, the scaffold ability on cell attachment, proliferation, and morphology was observed on a phase-contrast microscope (Figure. 9.16). The relative density of cells in the PCPP3/PCL scaffolds, TSS-HAP, and TSS-HAP1 to TSS-HAP4 coated PCPP3/PCL scaffold (60 µg/mL) were compared for 12 h, 24 h. Cell morphology and proliferation rate were similar for the PCPP/PCL scaffold and HAP alone. In TSS-HAP3 and TSS-HAP4 coated PCPP3/PCL scaffold shows the increases in cell proliferation rate and the elliptical cell morphology indicates the strong adhesion of cells. Comparing with other samples, TSS-HAP3 and TSS-HAP4 coated PCPP3/PCL scaffold exhibited a significant difference.

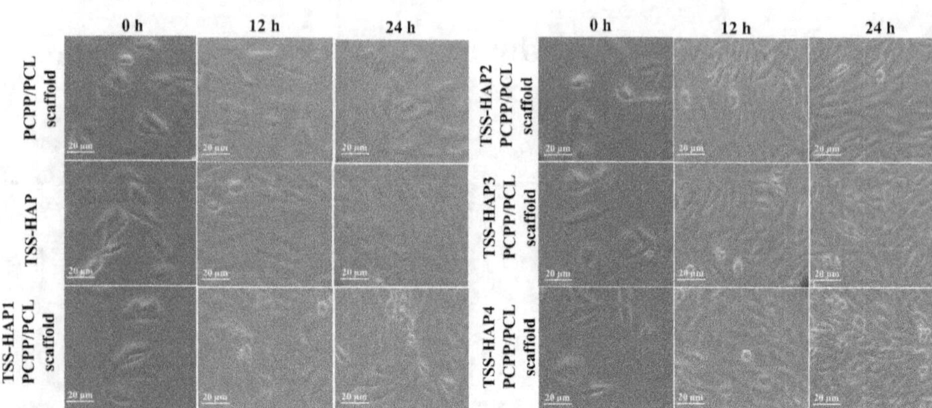

Figure. 9.16. Visualization of MG-63 cells morphology after treatment with TSS-HAP, PCPP3/PCL scaffold, TSS-HAP1, TSS-HAP2, TSS-HAP3, TSS-HAP4 coated PCPP3/PCL scaffold for 0 h, 12 h, 24 h by phase contrast microscope.

9.6.12. Biocompatibility analysis of scaffold via flow cytometry

HAP coated scaffold biocompatible effect on MG-63 cells was measured by flow cytometry by gating the viable, and dead cells. Plain scaffolds have exhibited the 0.52 % of dead cells and TSS-HAP coated scaffold shows 0.66 % of dead cells (Figure. 9.17. A,B). In figure. 9.17. upper right, green color quadrants denote the viable cells population and red color quadrants denote the dead cell population which binds with PI stain. Compared with the positive control, the scaffold and TSS-HAP coated scaffold exhibited higher biocompatibility (Figure. 9.17. C-F).

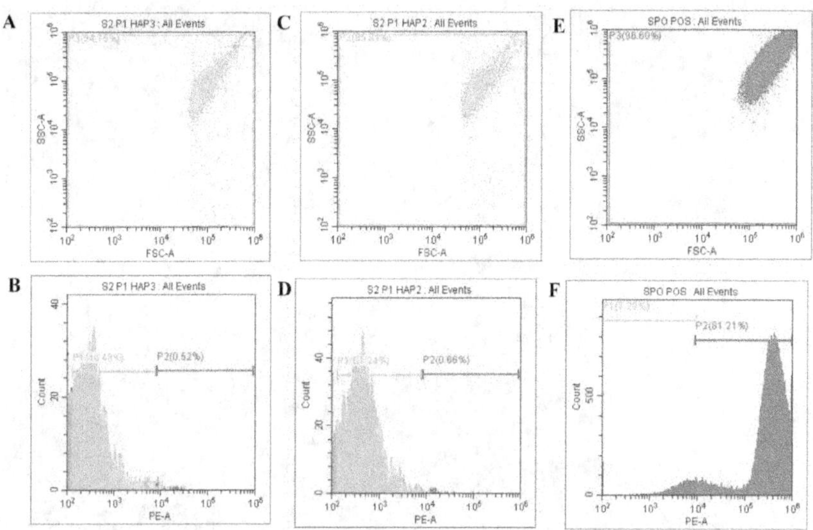

Figure. 9.17. (A,B) PCPP3/PCL scaffold biocompatible effect on MG-63 cells (C,D)TSS-HAP4 coated PCPP3/PCL scaffold biocompatible effect on MG-63 cells (E,F) Positive control cells biocompatible effect measured by flow cytometry.

9.6.13. SEM analysis

The TSS-HAP4 coated PCPP3/PCL scaffold itself has a significant effect on the MG-63 cells attachment which represents in Figure.9.18. A-D. Scaffold consist of the rounded structure of MG-63 cells over the surface of the scaffold. It indicates the scaffolds provide a framework, induce support for the cell's attachment, proliferate and act as an ECM. The surface topography, porosity, interconnectivity, mechanical strength, and functional groups are significantly influenced cell attachment and growth. TSS-HAP bioactivity increases the scaffold's regenerative applications like adhesion, differentiation, and growth.

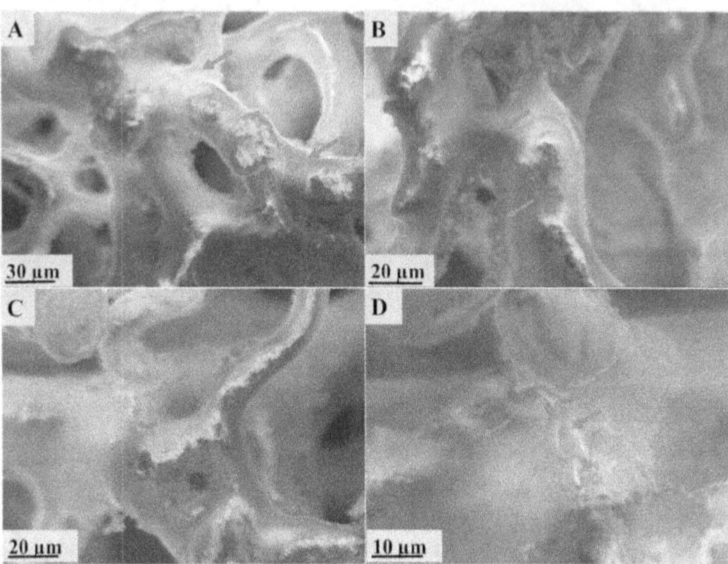

Figure. 9.18. SEM image represents the MG-63 cells growth, adhesion over the TSS-HAP4 coated PCPP3/PCL scaffold (red arrows- cells).

9.6.14. Antibacterial activity of scaffold

The antibacterial activity of PCPP3/PCL, TSS-HAP4, TSS-HAP4-PCPP3/PCL were evaluated against the *Staphylococcus aureus, Escherichia coli* represent in Figure. 9.19. A,B. PCPP3/PCL shows no antibacterial activity against bacterial culture. TSS-HAP4 and TSS-HAP4-PCPP3/PCL shows effective antibacterial effect in low concentration. In 30 µg/mL concentration, it shows better inhibition values were indicating the quality of material. The antibacterial efficiency helps in control the biofilm formation on the scaffold without any chemical agent.

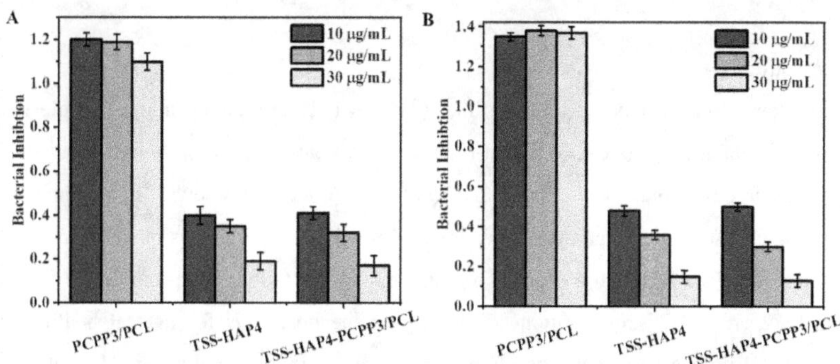

Figure. 9.19. Antiabcterial effect of PCPP3/PCL, TSS-HAP4, TSS-HAP4-PCPP3/PCL (A) *Staphylococcus aureus* **(B)** *Escherichia coli.*

9.7. Conclusion

The surface coating of PCPP3/PCL scaffolds with snail shells HAP was successfully demonstrated. Incorporation of HAP enhanced the scaffold's compressive strength and bioactivity. HAP particles were evenly dispersed in the scaffold which was confirm by SEM. After scaffold incubation in SBF buffer, it induces to form a mineral precipitates. The biomineralization of the scaffolds helps in improvement of bioactivity and cell proliferation. The bone-like apatite formation over the scaffold provide surface for the osteo-cell adhesion.

10.1. Introduction

Metals-based orthopedic implants like titanium (Ti) or titanium alloys, stainless steel 316L, cobalt-based metal alloys (Co-Cr), nickel-titanium, and magnesium were widely used. It is normally designed as fracture/bone plates, hip nails, joint caps, pins, and screws. It has higher mechanical strength, tensile strength, wear resistance, fracture toughness, and corrosion resistance. The implant was used to hold the bones in a specific region, support the tissue, and promote the osteointegration process [272,273]. Additionally, it needs to trigger the host cells for natural healing mechanism. Majorly, metal implant contacts to the human body fluids, there is a chance of lose its efficiency, in the ability to form a chemical bond between the bone tissue. Further, the higher release of metal ions leads to a negative effect on the organs [274,275]. The bioactive, surface functionalization, a corrosion-resistant metal implant is required to overcome the issue. Coating of bioactive nanomaterial on metal implants is considered to be a promising technique to enhance the biocompatibility, functionality, and increase the metal-tissue interactions [276,277]. Majorly, EPD is an extensively used coating method to deposit bioactive nanomaterial on metallic substrates. EPD is a simple technique, required lower-cost equipment for coating for even thick metal substrates with complex structures. Its processing rate was higher, and possible to control the coating thickness and microstructure by simple adjustment of voltage potential and time [278,279].

Halloysite nanotubes (HNT) is $Al_2Si_2O_5$ $(OH)_4.nH_2O$, the structure possesses hollow cylinders contains multiple-rolled layers formed with octahedral alumina plates. It has an active Al-OH group on the inner surface, and a tetrahedral (Si-O-Si) siloxane layer on the outer surface [280,281]. In that, HNT with a tubular structure able to load various components like bioactive agents, drugs. It can be entrapped into the inner lumen or void space of the multilayered HNT. Bioactive materials from marine origin are siliceous in nature and rich source of biomolecules, bioceramics, and rare bio-components, etc. In that, marine sponge (MS) a soft-bodied component consists of organic/inorganic compounds. Marine sponge *Callyspongia diffusa* (CS) contains sesterterpene, triterpenes metabolites which have antibacterial, anti-inflammatory, and bone regeneration process [282,283].

10.2. Previous work

There lesser report was available regarding the combination of nano clay for EPD coatings. HNT osteointegration and regeneration ability were enhanced by the entrapment of bioactive substances into the lumen [284,285]. Additionally, the retention and release of the active substance confirm the HNT well-appropriate natural nano-containers for macromolecular delivery. Marine sponge *Callyspongia diffusa* contains sesterterpene, triterpenes metabolites which have antibacterial, anti-inflammatory, and bone regeneration processes [286]. It was also reported for its cell proliferation factor, cell adhesion, and differentiation [287,288] It also a collective form of more trace elements, such as S, Zn, Mg, Mo, Fe, K and Ca naturally.

10.3. Novelty

HNT mineral was used to loading the *Callyspongia diffusa* MS and the surface of HNT was modified with PCPP to enhance the release of the bioactive component. PCPP enhances the mechanical and flame retardant properties [289]. It also controls the permeability of oxygen, water and subsequently improves the barrier property of coating material against corrosive species. [290].

10.4. Objective

In this work, HNT was loaded with a marine sponge, followed by coated with PCPP polymer via APTES (3-Aminopropyl)triethoxysilane). Silane-coupling agents have majorly used substances for surface alteration of HNTs. The APTES, a silane agent which helps in the modification of HNT charge on the surface, and as well as making stable chemical linkage of PCPP polymers on the HNT surface. Finally prepared CD-HNT-PCPP nanocomposite was deposited on Ti-6Al-4V screw through EPD. The coated implant was characterized for mechanical property, corrosion resistance, and morphology. *In vitro* release, biomineralization, cell adhesion, and proliferation were evaluated by MG-63 cells.

10.5. Experimental techniques
10.5.1. Preparation of MS loaded HNT

Marine sponge *CS* was ground into fine powder. The MS whole powder of 50 mg was dissolved in 1 mL of ethanol to form a solution. Additionally, 25 mg of halloysite was added to the above solution and mixed to get a homogeneous solution. The suspension was subjected to sonication for 10-15 min, then it was placed in a vacuum for 4 cycles. Then, the sample was washed with ethanol to remove the unbound CD. Further, the sample was dried in a vacuum desiccator for further use [291].

10.5.2. Functionalization of MS loaded HNT with APTES

The functionalization of the APTES on external surfaces of HNT was widely reported in several applications [292]. Initially, 1 g of MS-HNT was added in 25 mL of toluene, sonicated for 2 h. Then, 1 mL TEA and 2 mL APTES were added to the suspension and refluxed at 80 °C for 24 h under a nitrogen atmosphere. The resulting residue was washed under toluene and distilled water three times to eliminate the unreacted APTES. Further, MS-HNT-APTES was separated by centrifugation at 4000 rpm for 20 min and dried in an oven.

10.5.3. Fabrication of PCPP grafted MS loaded HNT

PCPP polymer was coated on MS loaded HNT using APTES which form a chemical linkage with PCPP. HNT-APTES of 0.2 g, 0.67 g of EDC, NHS were added to the water (20 mL). The solution mixture was kept in magnetic stirring for 2 h and 0.4 g of PCPP was added to the solution, stirred for 12 h at room temperature. The obtained MS-HNT-PCPP was centrifuged at 10,000 rpm for 20 min. Then washed with distilled water to eliminate the unbounded PCPP.

10.5.4. Titanium specimen surface pre-treatment

Ti-6Al-4V ELI screw materials of length- 5 mm and diameter-3 mm were used. The screw was treated with grit 600 grinding paper, rinsed in acetone, and water in bath sonication for 10 min to remove surface residues. Surface etching treatment was performed with HF (30 %) /HNO$_3$ (65 %) (2/20 % V, respectively) for 1.5 min, and then screw washed with water. The pre-treated screw was used for further electrodeposition coating of HNT nanocomposites.

10.5.5. Electrophoretic deposition

Electrophoretic coatings attained using a DC power supply (Ablab, L1282, 1–128 V) in a conventional two-electrode cell. The cathode was fixed as a Ti-6Al-4V screw, and the anode as a platinum electrode was kept at a distance of approximately 1 cm. For each experiment, 15 mL dispersion was used, and the coating was performed by applying different potentials (10–30 V) for 2 min. After deposition, the samples left to dry at ambient conditions.

10.5.6. Corrosion resistance behavior

Corrosion resistance behavior of HNT nanocomposites coating was assessed with changes in corrosion current density and potentials. The study conducted using three-electrode cells, and before the study, samples were immersed in SBF with pH 7.4 at 37 °C. The potentiodynamic polarization study was started after reaching the stable open circuit potential. Measurement was conducted out by increasing the potential from −0.4 VSCE at a scan rate of 1 mV s^{-1}.

10.5.7. Bioactive agent release studies

MS-PCPP-HNT coated titanium screws were immersed in PBS (pH 7.4) at 37 °C and kept in a humidity chamber for 8 days. Infrequent time, the buffer was replaced by a fresh PBS buffer. To investigate the MS release, samples solutions were measured at λ-max-220 nm via an HPLC set, which was manufactured by Agilent technologies, 1200 infinity series.

10.6. Results and discussion

The marine invertebrates can support bone formation and are composed of $CaCO_3$ skeleton, proteins, spongin, and siliceous spicules. In this study, CS marine sponge's whole vital components were loaded into HNT, further, it was coated with PCPP polymer, and finally, MS-HNT-PCPP was deposited on Ti-6Al-4V screw via EPD.

10.6.1. Charecterization of marine sponge-loaded HNT, polymer coated HNT
(i) FTIR analysis

The FTIR spectra of different concentrations of marine sponge-loaded HNT vibrational frequencies shift represent in Figure.10.1. Halloysite has peaked at 560, 913, 1039, 3694 and 3630 cm⁻¹ indicates the vibrations of Al–O–Si, inner Si–O stretching, and -OH of the interlayer (Figure.10.2. A). The MS of different concentrations loaded HNT exhibited major adsorption of HNT. The disappearance of vibration peaks 3694 and 3630 cm^{-1} (-OH stretching on the inner surface) indicated the packing of the MS in the HNT (Figure.10.1. A, B, C).

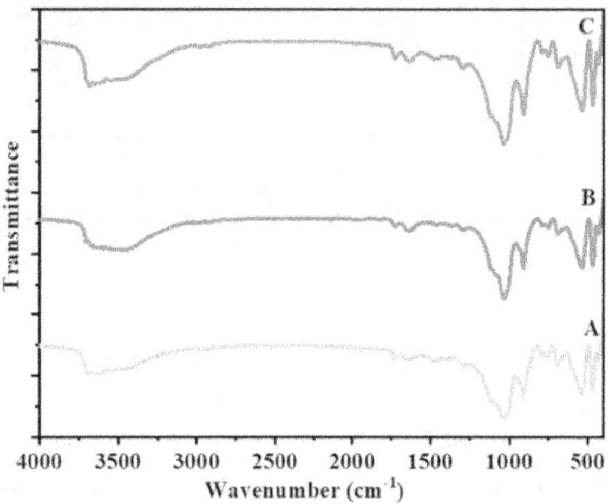

Figure.10.1. FTIR micrographs of (A) 100 mg marine sponge loaded halloysite (B) 200 mg marine sponge loaded halloysite (C) 300 mg marine sponge loaded halloysite.

In Figure.10.2.B, APTES functionalized HNT forms a new peak at 2941, 1369 cm^{-1} related to the CH$_2$ stretching and bending vibrations. The broad peak at 3440 cm^{-1} represents the amine group and peak 1072 cm^{-1} indicates the C–N bonding of APTES. The band 3695 cm^{-1} is related to the -OH group stretching of the inner-surface of HNT. It denotes the APTES functionalization on the outer layer of HNT without affecting the hydroxyl group at the inner surface of HNT. PCPP coated HNT shows the new absorption peaks at 1640 and 1580 cm^{-1} indicates the carboxylic, amine group of PCPP polymer (Figure.10.2.C). FTIR results confirm the PCPP coated on the outer layer of HNT with the help of APTES.

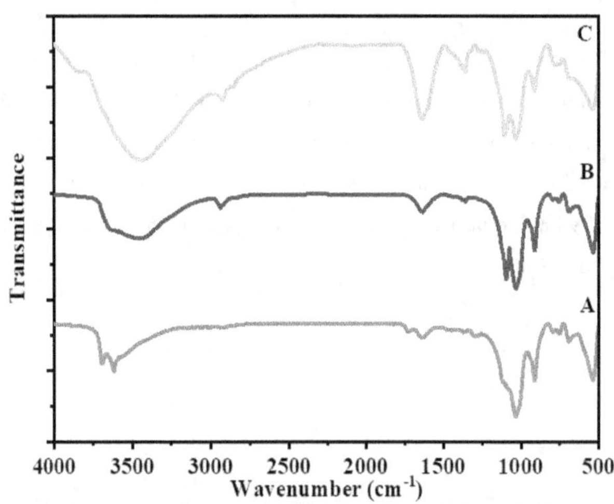

Figure.10.2. FTIR analysis of (A) HNT (B) AEAPS functionalized nanoclay (C) Polymer coated halloysite.

(ii) XRD pattern

XRD patterns of HNT, APTES-HNT, marine sponge-loaded HNT, and PCPP-HNT are shown in Figure.10.3, 10.4. After loading different concentrations of MS into the HNT, XRD represents the unchanged structure. APTES-HNT and marine sponge-loaded HNT have a basal peak at 2θ-12.24° corresponds to a (001) plane in the nanotube. Other peaks like 2θ-20.12, 25.21 which is indexed to the plane of (110), (002). The observed diffraction peaks are indexed to the halloysite (JCPDS 29- 1487) and confirm the tubular structure of the samples (Figure.10.3. A-B, 10.4. A, B). Also indicates the crystalline structure of the HNTs was not affected during the functionalization/loading processes.

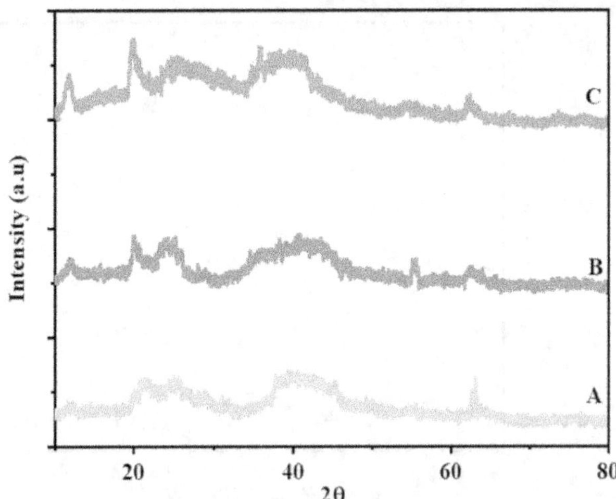

Figure.10.3. XRD pattern of (A) 100 mg marine sponge loaded halloysite (B) 200 mg marine sponge loaded halloysite (C) 300 mg marine sponge loaded halloysite.

After APTES functionalization and polymer grafting, HNT structure remains the same. PCPP was grafted onto the APTES amine groups which present on the outer surface of HNT. In PCPP-HNT, a new peak was formed at 2θ-29.18° which belongs to the PCPP diffraction peak and it indicates the existence of PCPP in HNT (Figure.10.4. C).

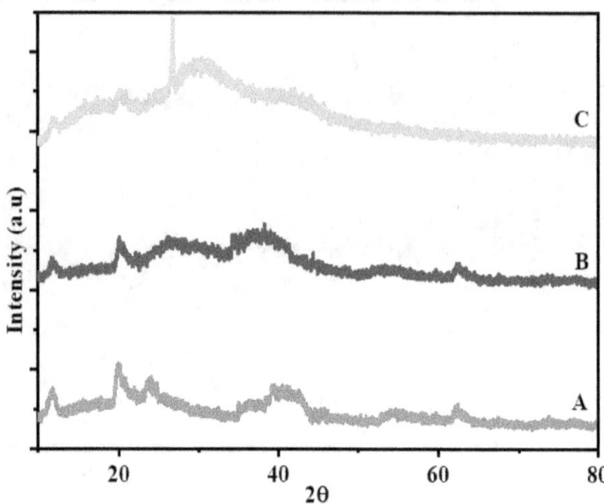

Figure. 10.4. XRD pattern of (A) HNT (B)AEAPS functionalized nanoclay (C) Polymer coated halloysite.

(iii) TGA studies

TGA studies show the thermal stability and degradation behavior of the pure HNT, and PCPP-HNT (Figure.10.5). According to the attained result, the weight loss of nano clay occurs at two steps. The first weight loss is around 250° C, corresponding to the dehydration process, removal of water from the interlayer. The second weight loss occurs around 450° C with a weight loss of 5 % attributed to the decomposition. PCPP-HNT steady weight loss occurs around 450° C, with a 5 % loss. Both follow the similar weight loss, the major variation was due to the decomposition of aminopropyl, PCPP grafted to the HNTs. From these TGA analyses, it can be concluded that the PCPP grafting was not influenced the thermal stability of HNT [293].

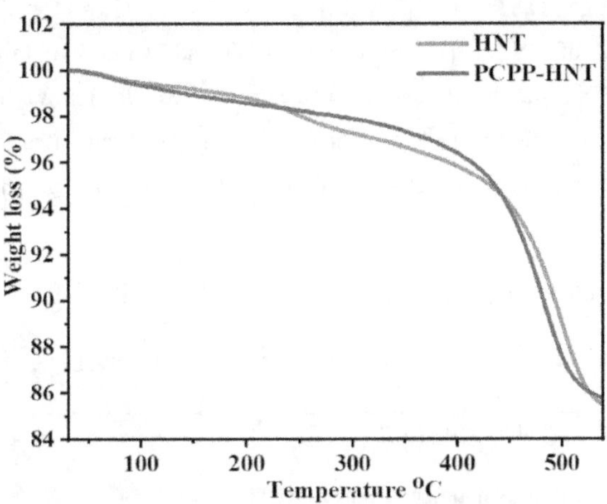

Figure.10.5. TGA analysis of nanoclay and PCPP-HNT.

(iv) Zeta potential studies

The change in zeta potential value of the HNT after APTES functionalization and polymer coating denotes the successful surface modification (Table.10.1). HNT forms a negative zeta value of -30.46 mV and it possesses a negative zeta value after loading with a MS of different concentrations. After conjugation with APTES, the external surface charge changes into positive +42.12 mV. Further PCPP coating alters the surface charge into − 45.24 mV and it confirms the polymer coating on the HNT and the stability of nanomaterial.

Table.10.1. Zeta potential value of HNT, MS-HNT, PCPP-HNT.

	Zeta value (mV)
HNT	-30.46
MS-HNT	-33.52
APTES-HNT	+42.12
PCPP-HNT	− 45.24

10.6.2. MS Loading of HNT

The loading of MS was performed in the HNT and PCPP-HNT. In that, the amount of MS loaded in PCPP-HNT was higher and the percentage of MS loaded was represent in Table.10.2. Marine sponges have sufficient electrostatic interaction with HNT, increases adsorption of MS through surface roughness property. PCPP-HNT showed effective drug loading and it used for the release studies.

Table.10.2. MS loading, encapsulation in HNT and PCPP-HNT.

	HNT	PCPP-HNT
Loading efficiency	6.76	23.22
Encapsulation efficiency	10.02	29.41

10.6.3. MS release from HNT in different pH condition

The experiments of MS release from PCPP-HNT were carried out in different pH conditions by the dialysis bag method (Figure.10.6). Initially, PCPP-HNT exhibited the burst release in both pH due to the rapid dissolution of HNT in the buffer. After 20 h, there was 26 % (pH 7.4), 30 % (pH 5.0) release of MS was observed. Further, release from HNT lumen was slower for a longer period. After 100 h, 58 % (pH 7.4), 62 % (pH 5.0) denotes the release attained the steady-state. PCPP-HNT release of MS was in a more controlled manner and polymer chain controls the burst release of marine sponge. Major opening/cavities of HNT were covered by a polymer chain. The release of HNTs was faster in the acidic condition compared to the neutral medium due to the faster erosion of the polymer chain. However, the MS release was done in a more controlled way by PCPP-HNT.

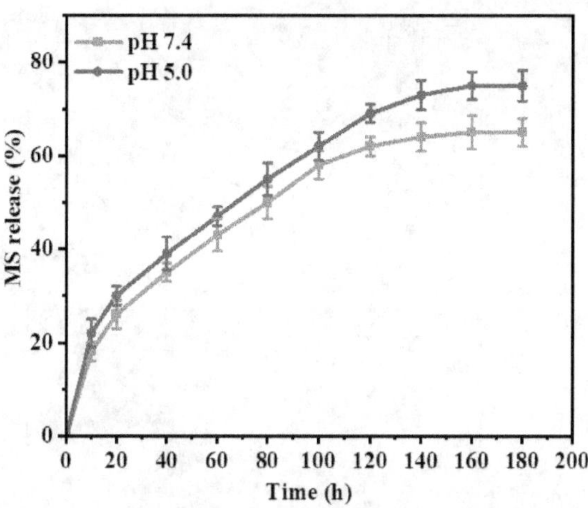

Figure. 10.6. MS release from PCPP functionalized HNT under different pH condition.

10.6.4. SEM analysis

SEM analysis showed the surface morphology and shape of the PCPP coated HNT with lumens size ranging from 15 to 20 nm (Figure.10.7). After surface modification of the halloysite with AEAPS and PCPP coating has not altered the tubular structure of HNT. Figure. 10.7.D represents the marine sponge-loaded PCPP-HNT with a more rugged surface. The outer layer of HNT was thickened by polymeric bonding, the tubular structure is not damaged during the marine sponge loading and conjugation process. PCPP is anchored on HNT by the covalent linkage between the amine and carboxylate groups. In higher magnification, it shows some polymer crystals onto the outer surface of HNT [294]. The thick coating of polymer over the halloysite tubes, provide dispersible and less aggregated material.

Figure.10.7. SEM image represents PCPP functionalized MS-HNT in different magnification.

10.6.5. EPD and coating characterization of PCPP-HNT on Ti-6Al-4V screw

After the marine sponge loaded in PCPP-HNT, the stable PCPP-HNT-MS was studied their EPD on Ti screw. To identify the optimum condition, we have taken three different concentrations of PCPP-HNT-MS (0.3, 0.6 g) and potential (10, 20, 30 V). The roughness, coating thickness, and deposit weight of various groups were represented in Table.10.3. In 0.3 g PCPP-HNT-MS, the roughness of the samples was increased with an increasing voltage potential. Similar results for the 0.6 g concentration of PCPP-HNT-MS which provides the higher roughness of 0.94 μm for 30 V. For all groups, the roughness was higher than that of a reference (Sa = 0.13 μm). The surface roughness, the thickness of implant substrate is a vital factor for the adhesion of osteoblast cells. Increases in surface roughness improve cell adhesion and proliferation. The PCPP-HNT-MS coating thickness represents the increase of voltage potential and HNT content leads to a raise of coating thickness. Especially, HNT concentration influenced the coating thickness compared to the voltage potential. In 0.3 g concentration, lower potential shows the lesser thickness, which indicates the improper migration of the HNT during the EPD process. The highest coating thickness was exhibited for the 0.6 g with 30 V potential and higher coating thickness may also cause the formation of crack, reduction in mechanical, adhesion properties. The deposition weight was increase based on the enhancement of HNT content and voltage potential. Another research also revealed the same result with increases in the applied voltage will increase the deposition of HAP particles significantly [295].

Table. 10.3. Thickness and deposit weight of PCPP-HNT-MS on Ti-6Al-4V screw in different condition.

PCPP-HNT-MS (g) in 100 mL of organic solvent	EPD voltage [V]	Sa parameter (µm)	Thickness (µm)	Deposit weight (mg)
	10	0.58	2.01	96
0.3	20	0.63	2.52	102
	30	0.69	2.92	119
	10	0.71	3.09	130
0.6	20	0.78	3.66	225
	30	0.94	7.24	350

10.6.6. Corrosion resistance studies

The corrosion studies were performed on various group coating and reference Ti screw without coating (Figure. 10.8). The values of corrosion current density and corrosion potential indicate the corrosion resistance of the material. All the groups have higher positive corrosion potential compared to the uncoated reference substrate. In 0.3 g, concentration, three groups showed the lower corrosion potential compared to the 0.6 g concentration. This result was due to the lower thickness of the coating, formation of corrosion channels which help localization of SBF and enhances the electrochemical dissolution. The coating of 0.6 g HNT content with 20 and 30 V potential shows a higher corrosion potential. The higher anti-corrosion resistance was due to the thick, and dense surface coating. Based on all groups coating properties, 0.6 g PCPP-HNT-MS with 30 V shows higher roughness and thickness of the coating on the substrate. This leads to the formation of cracks, and loss of mechanical strength of the coating. So, 0.6 g PCPP-HNT-MS with 20 V was considered to be the optimum level.

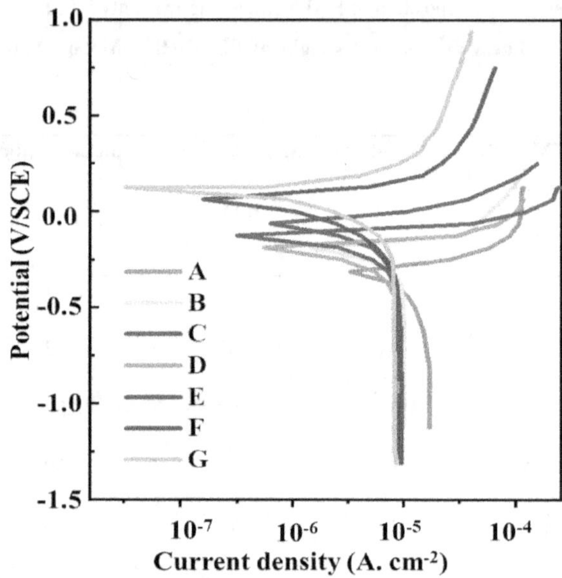

Figure. 10.8. Potentiodynamic polarization curves of (A) Ti-6Al-4V screw PCPP-HNT-MS/voltage varied samples (B) 0.3/10 (C) 0.3/20 (D) 0.3/30 (E) 0.6/10 (F) 0.6/20 (G) 0.6/30 coated Ti-6Al-4V screw.

10.6.7. SEM image of PCPP-HNT-MS coatings on Ti-6Al-4V screw

SEM micrographs of PCPP-HNT-MS coatings on Ti screw are shown in Figure. 10.9. A-L, at different magnification. In lower magnification reveals that the substrate was entirely covered by nanocomposites. It deposited homogeneously, the coating was evenly deposited on every part of the screw. Thread of screw which consists of a white layer indicates the coating was good without any space. EPD exhibited dense coating, interconnected network, and the reinforced halloysite particles on screw without any crack. This will provide better corrosion resistance to the underlying Ti screw. In higher magnification, nanomaterial denotes agglomerated, crystal-like structure in most of the area, but it distributed over the entire surface of the substrate. It also shows the elongated, tubular structured nanomaterial [296].

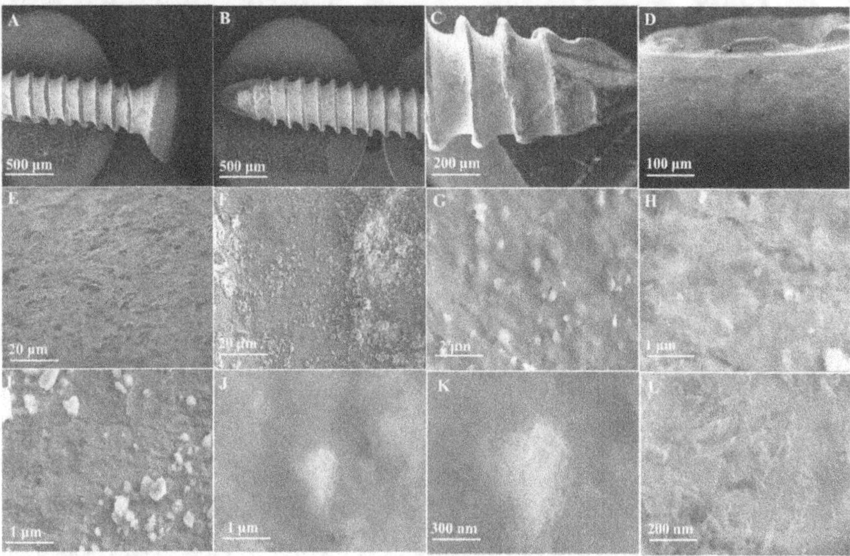

Figure. 10.9. SEM image of PCPP functionalized MS-HNT coated titanium screw surface via EPD at different magnification.

10.6.8. SEM image of PCPP-HNT-MS apatite

The biomineralization studies were performed incubation of PCPP-HNT-MS in SBF for 21 days at 37 °C. PCPP-HNT-MS before the immersion into the SBF was represented in Figure.10.10. A-F. It denotes the nanomaterial was homogenously distributed, with complex tubular geometry, polymeric layer on the outer surface. SEM image of PCPP-HNT-MS after 21 days immersion in SBF at 37 °C. It forms agglomerated pellet-like clusters, aggregated with an apatite crystal on the surface of materials that resembles the biomineralization of PCPP-HNT-MS. Tubular mineralized apatite nanocrystals were formed with help of minerals in the marine sponge which has a significant impact on the crystallization under SBF conditions. The bioactive HNT serves as nucleation sites for apatite formation. The formation of apatite crystals *in vitro* conditions suggests the HNT ability on osteoblast cell interaction [297].

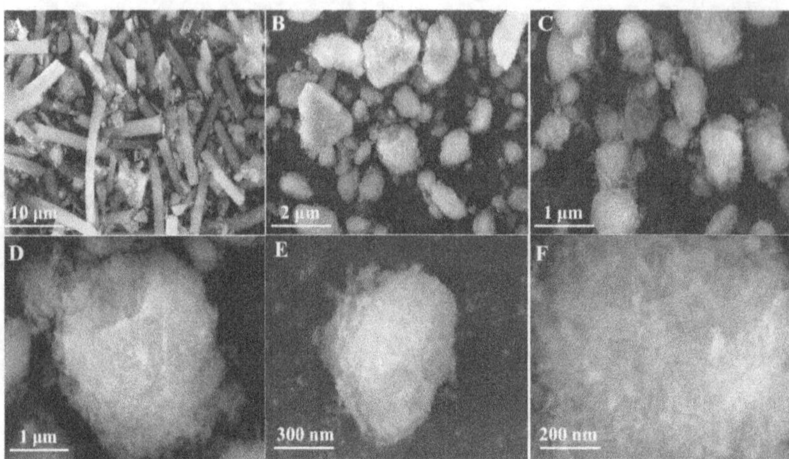

Figure.10.10. SEM image of (A) PCPP functionalized HNT (B-F) Biomineralization PCPP functionalized HNT immersed in SBF for 21 days which visualized in SEM at different magnification.

EDS elemental analysis was performed to identify the composition of the coatings before and after the SBF treatment. Figure10.11, shows the EDS spectra of PCPP-HNT-MS before SBF treatment, consist of Si, Al, O, C, and N. This confirms it consists of HNT and organic compounds of the polymer, APTES.

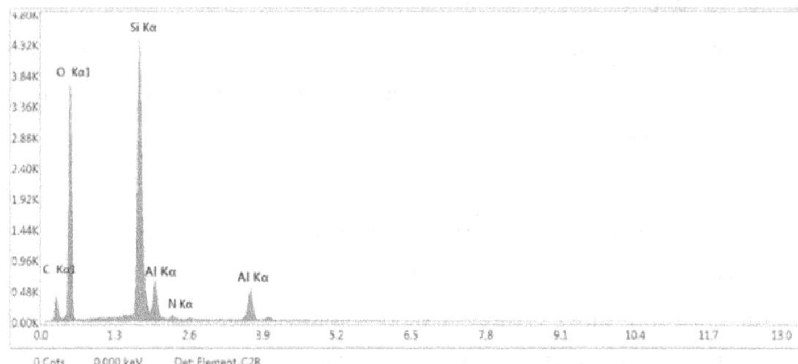

Figure.10.11. EDS analysis of PCPP-HNT-MS before SBF treatment.

After SBF treatment, the spectra indicate the formation of biomineralization, where it consists of minerals such as Zn^{2+}, Mg^{2+}, Fe, Mo, P, Ca. It forms crystallized nonstoichiometric mineral-containing bonelike apatite on the HNT core (Figure10.12). The EDS spectrum shows the PCPP-HNT-MS coating can form mineral-enriched apatite under SBF conditions.

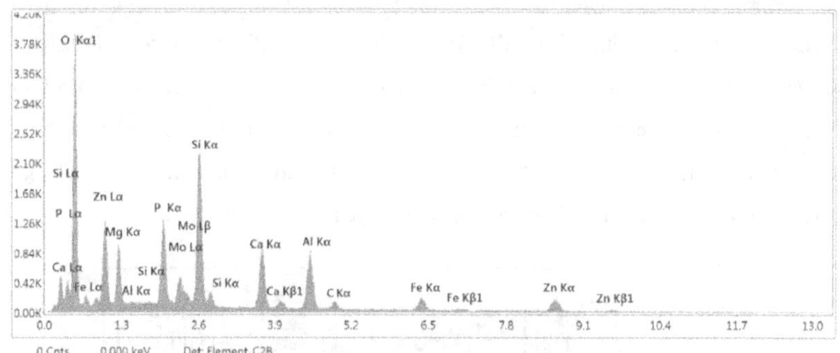

Figure.10.12. EDS analysis of PCPP-HNT-MS after SBF treatment.

Further elemental analysis was performed which shows the distribution of minerals such as Zn^{2+}, Mg^{2+}, Fe, Mo, P, Ca on the apatite (Figure. 10.13). These minerals have significantly induced crystallization, ions were adsorbed by HNT and helps in crystallization of the lattice.

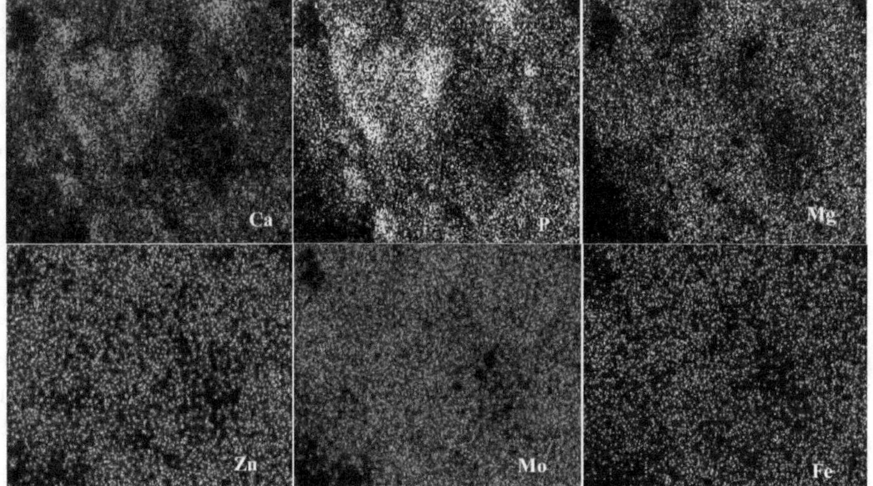

Figure.10.13. Elemental mapping of PCPP functionalized HNT-MS after 21 days of SBF treatment.

10.6.9. Cellular studies

(i) MG-63 cell proliferation

The assessment of nanomaterial cell proliferation ability is a crucial factor in the coating application. So, it was evaluated on MG-63 cell lines with different concentrations of HNT, PCPP-HNT, MS-PCPP-HNT (Figure. 10.14). There is an insignificant decrease of cell proliferation for the HNT of 100 µg/mL concentration. But, MG-63 osteoblast cells displayed higher cell viability for the PCPP-HNT-MS. Majorly it shows the lesser toxic in the concentration up to 100 µg/mL and also lesser reduction of cell viability based on the dose-dependent MS-PCPP-HNT.

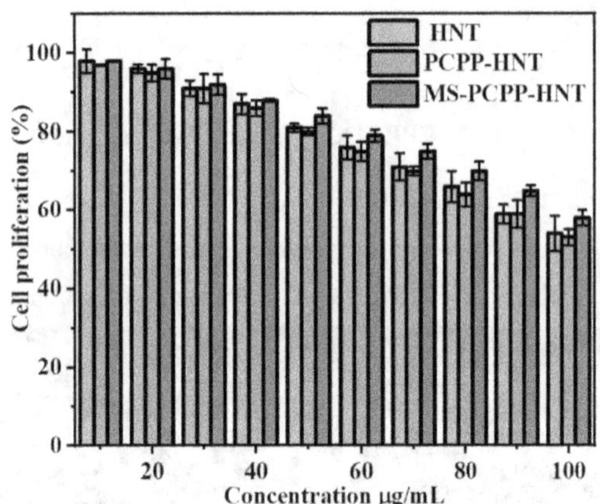

Figure. 10.14. Cell proliferation ability of HNT, PCPP-HNT, MS-PCPP-HNT on MG-63 cells.

(ii) ALP activity

The ALP activity measures the MS-PCPP-HNT nanomaterial osteoblastic effect on MG-63 cells. Figure. 10.15 represents the increases in alkaline phosphatase level, after treatment with HNT, PCPP-HNT, PCPP-HNT-MS. The HNT, PCPP-HNT exhibited the 1.5, 4 mM ALP level on the 21st day. PCPP-HNT-MS stimulate the highest amount (8 mM) of ALP on the 21st day, which was higher compared to the other two samples. This higher ALP activity induces osteogenic, cell differentiation, and colonization.

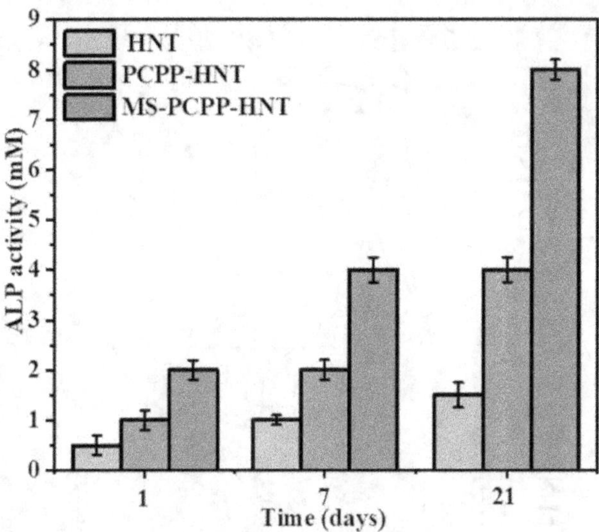

Figure.10.15. ALP activity of HNT, PCPP-HNT, MS-PCPP-HNT on MG-63 cells.

(iii) Clonogenic assay

Halloysite nanocomposite coated Ti screw treated with MG-63 cells can form colonies and it was evaluated by clonogenic assay. The colony formation rate was similar for HNT and PCPP-HNT for 7, 14, and 21 days (Figure.10.16). The rate of colony formation was increased for the PCPP-HNT-MS treatment. It forms high dense colonies on the 21st day which represent by the intense blue color.

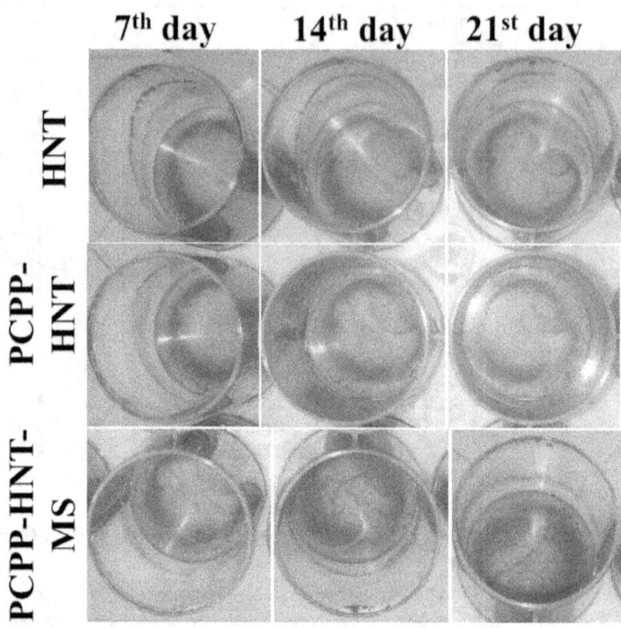

Figure.10.16. Clonogeny study of MG-63 cells after treatment with HNT, PCPP-HNT, MS-PCPP-HNT for 7, 14, 21 days.

(iv) MG-63 cells morophological studies

Figure.10.17 represents the morphological characteristics of MG-63 cells grown on the HNT nanocomposites. Phase-contrast images revealed the structure, proliferation, and adhesion ability of MG-63 cells after treatment. It also suggests all three samples of 60 µg/mL concentration show no toxigenic effect on the cells, but it improved the adhesion nature and proliferation of cells. Comparing with HNT, PCPP-HNT, PCPP-HNT-MS increases the cell proliferation rate. The HNT, PCPP-HNT have randomly dispersed cells with thin morphology, rounded structure, lesser colonies. In PCPP-HNT-MS, cells were highly dense, tightly packed, flat appearance, elongated, rough monolayer, and stellar-shaped.

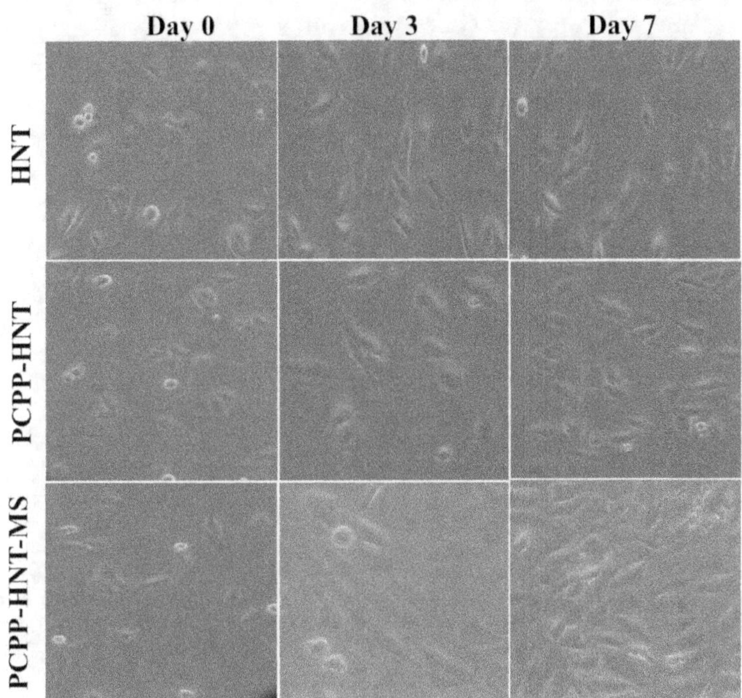

Figure. 10.17. Visualization of MG-63 cells morphology after treatment with HNT, PCPP-HNT, MS-PCPP-HNT for day-0, 3, and 7 by phase contrast microscope.

(v) Biocompatibility effect of PCPP-HNT-MS via flow cytometry

The biocompatible effect of HNT, PCPP-HNT, and PCPP-HNT-MS was evaluated on MG-63 and measured by gating the viable, and dead cells through flow cytometry. HNT has exhibited the 0.35 % of dead cells and PCPP-HNT shows 0.56 % of dead cells (Figure. 10.18. A-B). In Figure. 10.18. C-F, upper right, green color quadrants denote the viable cells population and red color quadrants denote the dead cell population which binds with PI stain. Compared with the HNT and PCPP-HNT, the PCPP-HNT-MS have exhibited higher biocompatibility.

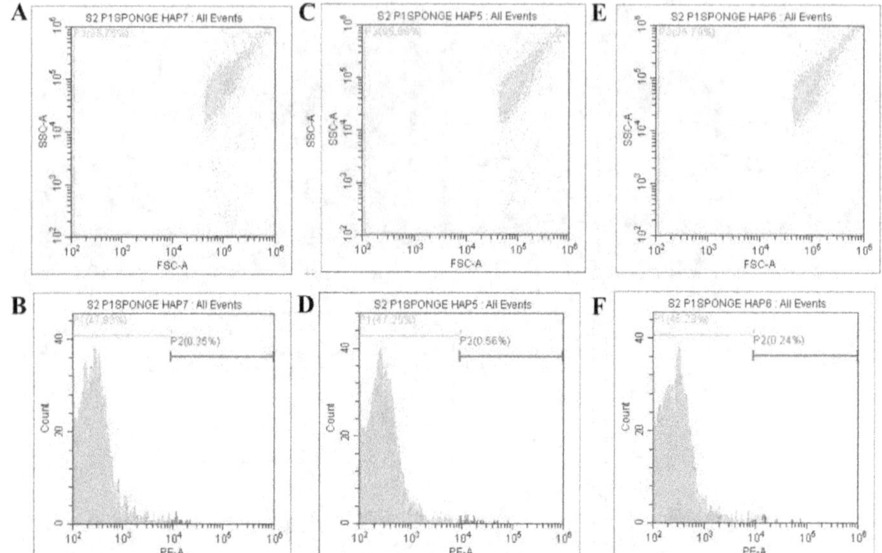

Figure.10.18. Flow cytometry for identification of viable/dead cells along with histograms peak shift of dead cells (A,B) HNT (C,D) PCPP-HNT (E,F) PCPP-HNT-MS on MG-63 cells.

10.6.10. MG-63 cells adhesion on MS-PCPP-HNT

The SEM analysis illustrates the morphology and adhesion characteristics of MG-63 cells grown on the HNT nanocomposites coated Ti screw. In Figure.10.19. A-D, SEM images revealed the rounded shape, flattened shape with high interconnectivity was observed. MG-63 cells attached and spread over the surface of nanocomposite and there was an increase in cell number. It was confirmed by more number of cell attachment processes can be seen in higher magnification and it confirms the strong attachment. The structure of the adhered cells was varied due to the properties of the nanocomposites like wettability, charge difference, and surface roughness [298].

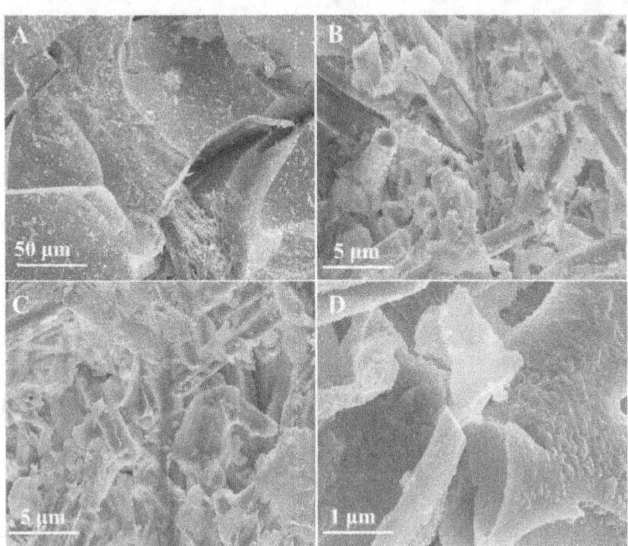

Figure.10.19. SEM image represents the MG-63 cells growth, adhesion over the MS-PCPP-HNT.

10.6.11. Antibacterial activity

In Figure.10.20. A, B, the antibacterial effect of the HNT, PCPP-HNT, MS-PCPP-HNT was performed against both *Staphylococcus aureus* and *Escherichia coli* in different concentration. There is no antibacterial effect on HNT and PCPP-HNT treatment in different concentrations. MS-PCPP-HNT show the antibacterial effect in both *S. aureus*, and *E. coli* at 10 µg/mL concentration. Further increase in concentration leads to higher inhibition of bacteria. Bioactive sponge consists of metabolites like sesterterpenes, triterpenes which have a high antibacterial effect, helping to address the globally challenging issue of regenerating infective bone tissues.

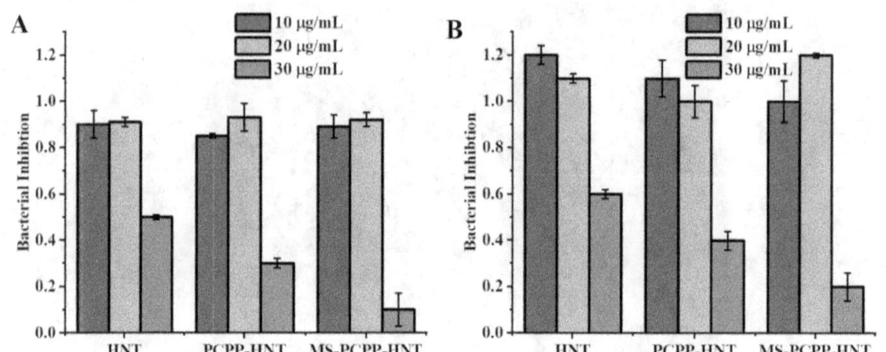

Figure.10.20. Antiabcterial effect of HNT, PCPP-HNT, MS-PCPP-HNT (A) *Staphylococcus aureus* **(B)** *Escherichia coli.*

10.7. Conclusion

The fabricated marine sponge loaded nanoclay consists of a unique source of minerals, trace elements, flavonoids, sesterterpenes, triterpenes, etc. The bioactive nanocomposites were coated on the Ti-6Al-4V screw via EPD and the coated components form a compact halloysite structure, evenly distributed on the surface. It formed strong attachment, mechanically stable with anti-corrosion property. It also induces the biomineralization to form bone-like apatite which helps in cell attachment and proliferation. Marine component coating possesses harmless with higher efficiency, and this could be further explored in the development of various forms of implant and coatings materials.

Summary and Future scope of the work

In this thesis, some challenges have been determined during the development of polymeric multifunctional nanomaterials. It also emphasizes the state-of-the-art of PCPP polymers, a different form of nanostructures preparative methods, difficulties concerning the desired properties, with diverse therapeutic application-cancer, TB, and bone regeneration. The enhanced superior properties of PCPP overcomes the limitation of conventional polymer drug delivery and resolve the many therapeutic defects. Poly(organo)phosphazenes polymer has a wide range of applications in the medical field. Consequently, it can fabricate into different forms based on the application and also increase therapeutic efficiency.

Chrysin and cisplatin anticancer drugs were loaded in the MLNPs system, a promising therapeutic strategy against oral cancer. Cisplatin and chrysin co-encapsulated in CCNPs and covered with alternate layers of PDCPP/ PDADMAC polymers. A strong cellular internalization exhibited by MLNPs in KB cells and synergistic effect of the dual drug-induced cell apoptosis, elevates ROS levels, mitochondrial membrane damage, and alters apoptotic protein expression. The combinational drug resulted in superior therapeutic effect in tumor-induced hamsters by a higher rate of tumor regression as compared to cisplatin loaded MLNPs.

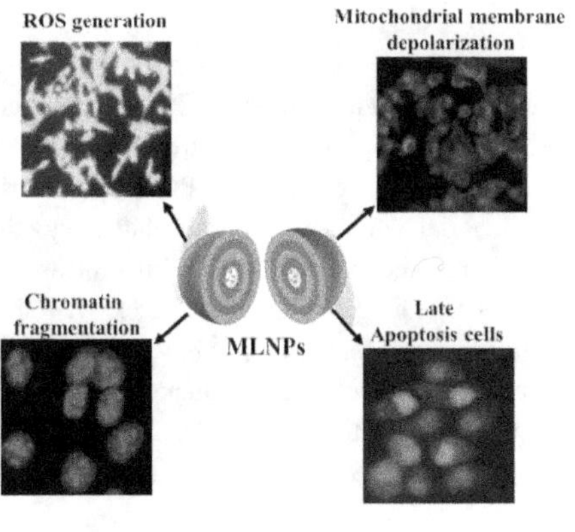

Hybrid polymer PCPP-PLA is the suitable new class of polymer with stimuli-responsive behavior. A hydrophilic - hydrophobic biodegradable hybrid polymer was formulated via PCPP (hydrophilic) and PLA, CA (hydrophobic). Hydrophobic drug PTX was successfully encapsulated with the hybrid polymer to form the nano range spherical carrier. The drug packed hybrid polymer provides a more sustained release of PTX drugs and shows effective anticancer effects against breast cancer.

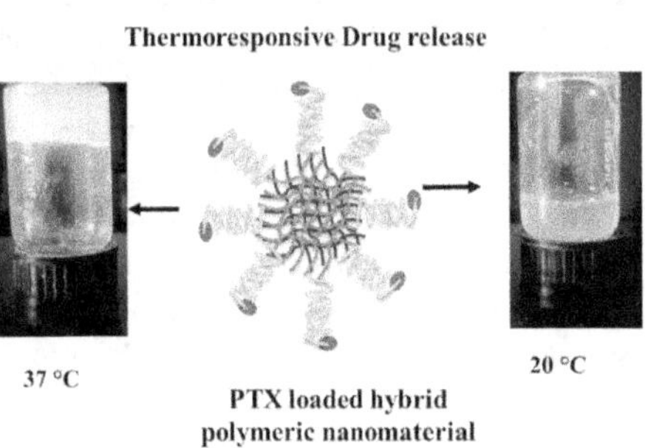

The CA functionalized polymeric nanomaterial delivers the drug to the targeted site and improves the therapeutic efficiency of PTX without systemic toxicity. The CA which has carboxyl groups at the end conjugated into PCPP using ethylenediamine and sodium cholate forms ionic interaction of PDADMAC. The co-polymeric nanomaterial was prepared and loaded with hydrophobic anticancer drug PTX. Targeting breast cancer cells by CA moiety was confirmed by efficient cellular uptake, intracellular release, and DNA damage of cells.

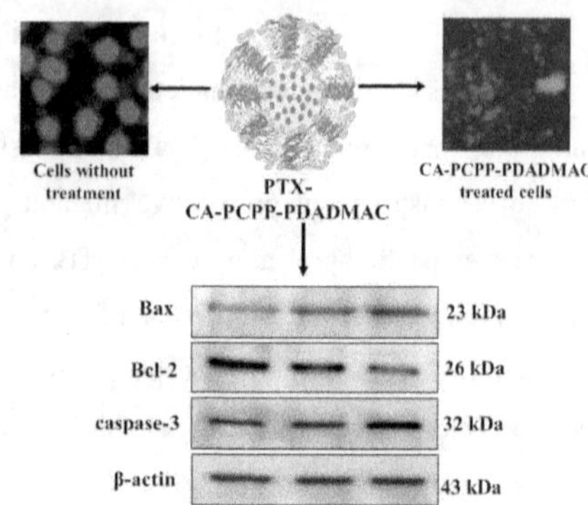

Active targeting CA conjugated nanomaterial loaded core-shell PHM nanofibers were prepared by coaxial electrospinning. PTX loaded into CA conjugated PCPP polymeric nanomaterial which were encapsulated into the nanofibers. The pH-responsive drug release of nanofibers exhibited intensely and ex *vivo* skin permeation of nanofibers shows high skin permeation. PTX eluted from nanofibers inhibits the growth of MCF-7 cancer cells through the generation of ROS and cell cycle arrest.

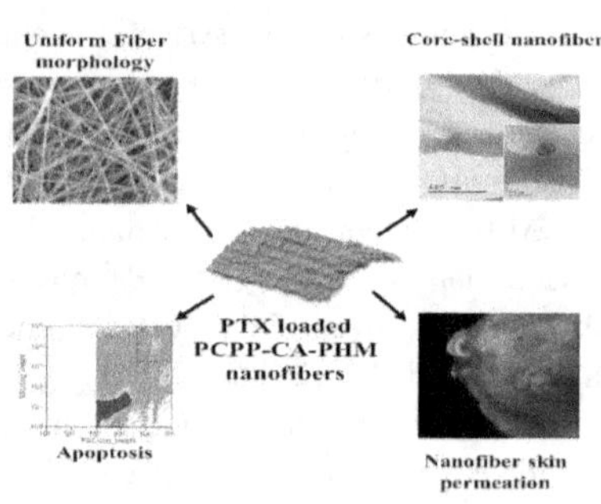

Arginine-g-PCPP was functionalized with liposomes for immunomodulating and intracellular delivery. Different liposomes formulation is utilized to identify the optimum sized liposomes and it was coated with arginine-g-PCCP. It shows endosomal pH-induced enhanced drug release inside the cell. Arginine activates the macrophage by increasing cell proliferation and more nitric oxide production. Liposomal delivery system with immune-stimulating ability is to deliver both drugs to the cytosol where *M. tuberculosis* survives.

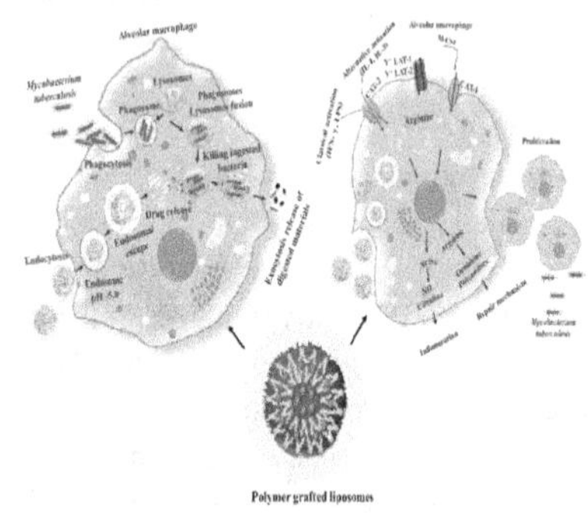

The all-in-one photoactive nanoagent composed with PCPP polymer (rGO-MoS$_2$-

PCPP-IZN-ICG) has effective antimycobacterial activity. Concretely, it not only exhibits the photothermal activity under NIR exposure but also works on singlet oxygen generation and antimycobacterial drug release. The bacterial membrane damage starts from the NIR-triggered nanomaterials which were systemically confirmed by *in vitro* studies. The utilization of rGO have a platform for MoS_2 functionalization, carries ICG and IZN drug. Combination effects accelerate the death of vulnerable TB bacteria which achieved by the single nanosystem.

HAP from the different snail shells has a variety of properties and was coated on PCPP/PCL by the dip-coating method. It formed a porous structure, interconnected network, and flake-like HAP structure. The prepared polymeric scaffold is a bone replacement material to promote bone formation. It provides mechanical support for adhesion of osteo-cells, integration, and adhesion.

A new approach for the coating Ti screw with ceramic nanocomposite with marine sponge consists of a unique source of flavonoids, sesterterpenes, triterpenes, etc. A PCPP polymer-coated nano clay-bioactive marine sponge was evenly distributed on Ti screw via EPD. The highly controlled coating of each component showed a compact flake-like structure, strong adhesive, mechanical, and anti-corrosion properties. Coated substrate forms bone-like apatite with high crystallinity, and cell proliferation, biocompatibility ability of coating was also high.

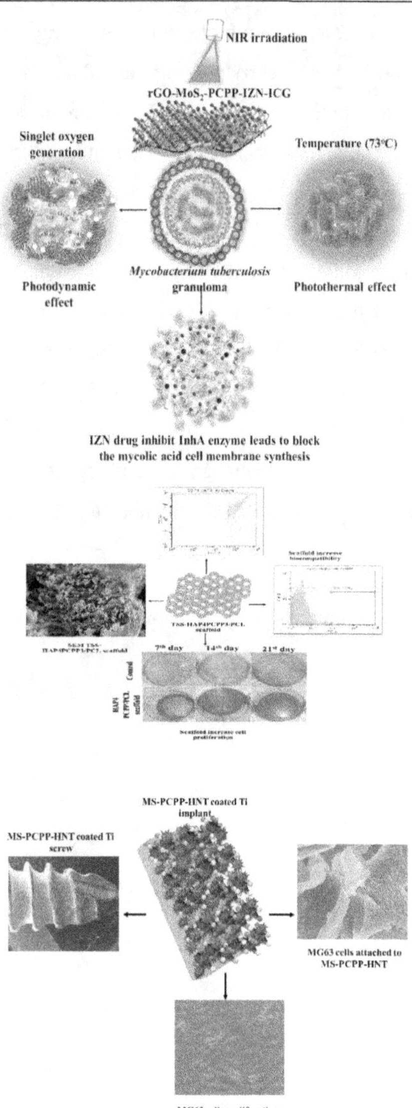

Overall, multiple anticancer drugs can be loaded into the poly[(organo) phosphazenes] multilayer nanocarrier system. It is the right option to deliver poorly soluble drugs, increases the bio-distribution, sustained release, reduced systemic toxicity. The formation of hybrid poly(organo)phosphazenes in a controlled manner to exhibit the pH and thermoresponsive property. Targeted drug delivery and transdermal drug delivery by multifunctional polymeric nanomaterial improve the therapeutic action of the drug. In addition to the drug delivery, antibacterial property alteration of the immune system against the TB bacteria is a promising approach. The combination of other therapeutic modalities with multifunctional nanomaterial demonstrated effective killing of TB bacteria specifically. Multiple factors affect the clinical success of bone implants, including osseointegration, and biocompatibility. The developed PCPP scaffold and polymeric coated Ti screw show good integration ability with bone cells. The incorporation of bioactive ceramic materials plays a major role in osteointegration and regeneration. The demand for nano-based biomaterials was high in the biomedical field and a nanomaterial with multifunctional property will convene the requirements in the future.

www.ingramcontent.com/pod-product-compliance
Lightning Source LLC
Chambersburg PA
CBHW071740150726
47998CB00005B/1727